TEACHING AND LEARNING A

Astronomy is taught in schools worldwide, but few schoolteachers have any background in astronomy or astronomy teaching, and available resources may be insufficient or non-existent. This volume highlights the many places for astronomy in the curriculum; relevant education research and "best practice"; strategies for pre-service and in-service teacher education; the use of the Internet and other technologies; and the role that planetariums, observatories, science centers, and organizations of professional and amateur astronomers can play. The special needs of developing countries, and other under-resourced areas, are also highlighted. The book concludes by addressing how the teaching and learning of astronomy can be improved worldwide. This valuable overview is based on papers and posters presented by experts at a Special Session of the International Astronomical Union.

JAY PASACHOFF is President of the International Astronomical Union's Commission on Astronomy Education and Development. He is also Director of the Hopkins Observatory and Chair of the Astronomy Department at Williams College, Williamstown, Massachusetts. In 2003 he received the Education Prize from the American Astronomical Society. His previous books include *A Field Guide to the Stars and Planets* and *The Complete Idiot's Guide to the Sun*.

JOHN PERCY is Professor of Astronomy and Astrophysics at the University of Toronto, and cross-appointed to the Ontario Institute for Studies in Education at the same university. His research program on variable stars and stellar evolution actively involves undergraduate students and skilled amateur astronomers. He has served as president of several scientific and educational organizations, including the International Astronomical Union Commission on Astronomy Education and Development, and the Astronomical Society of the Pacific. In 2003 he received the University of Toronto's Northrop Frye award "for exemplary linkage of teaching and research."

TEACHING AND LEARNING ASTRONOMY

Effective strategies for educators worldwide

JAY M. PASACHOFF
Williams College, Williamstown, Massachusetts, USA

JOHN R. PERCY
University of Toronto, Toronto, Ontario, Canada

CAMBRIDGE
UNIVERSITY PRESS

CAMBRIDGE UNIVERSITY PRESS
Cambridge, New York, Melbourne, Madrid, Cape Town, Singapore, São Paulo, Delhi

Cambridge University Press
The Edinburgh Building, Cambridge CB2 8RU, UK

Published in the United States of America by Cambridge University Press, New York

www.cambridge.org
Information on this title: www.cambridge.org/9780521115391

First published 2005
This digitally printed version 2009

A catalogue record for this publication is available from the British Library

ISBN 978-0-521-84262-4 hardback
ISBN 978-0-521-11539-1 paperback

Contents

Part X   Conclusions

Illustrations

Preface

This book is based on the proceeedings of a conference on "Effective Teaching and Learning of Astronomy" held on July 24–25, 2003, in Sydney, Australia, as part of the 25th General Assembly of the International Astronomical Union (IAU). It followed two previous IAU conferences on astronomy education, held in Williamstown, Massachusetts, USA, in 1988 and in London, UK, in 1996. The conference was organized within the framework of the work of IAU Commission 46 on Astronomy Education and Development. A major emphasis in the 1996 conference was the educational potential of the Internet, and robotic and remote telescopes, and it was heartening, in 2003, to see the growth and maturity of these technologies. But the basic challenges of astronomy education – especially at the school level – still remain.

The Organizing Committee for the conference consisted of: John Dunlop (New Zealand), Julieta Fierro (Mexico), Michèle Gerbaldi (France), Mary Kay Hemenway (USA), Syuzo Isobe (Japan), Barrie Jones (UK), Margarita Metaxa (Greece), Jayant Narlikar (India), Wayne Orchiston (Australia), Jay M. Pasachoff (USA), John R. Percy (Canada, Chair), Case Rijsdijk (South Africa), Rosa M. Ros (Spain), and Graeme White (Australia), with valuable help from the local organizing committees for the General Assembly, especially Nicholas Lomb, Sydney Observatory. Special thanks are due to Magda Stavinschi (Romania), whose proposal for an IAU Resolution with respect to astronomy education was a significant inspiration for this conference.

Over a hundred astronomers and educators from over two dozen countries attended or contributed to this conference. These included the presidents of the IAU, of the American Astronomical Society (USA), and the Royal Astronomical Society (UK), among others.

The theme of the conference was effective teaching and learning of astronomy at the school level, with emphasis on strategies that could be shown to be effective by research, assessment, or experience. The primary topics of the conference were addressed by 18 invited speakers from around the world. The questions, comments, and other discussion after these papers are recorded in this book. There were also general discussions of the place of astronomy in the school curriculum, and of future directions and initiatives in the effective teaching and learning of astronomy; these too are recorded here. In addition, there were several dozen contributed papers that were presented as posters at the conference. The abstracts of these have been woven into this book. Finally, there were a number of papers that were submitted to this conference by individuals who, in the end, were not able to attend. We have included brief summaries of many of these papers as well. A subsequent conference on Communication in Astronomy was held in Washington, DC, and led to the adoption of the "Washington Charter" for public outreach. We include the charter and some discussion of its relation to the goals of this book. We thank all the authors and contributors. This book is primarily their work.

We acknowledge the generous support of the IAU and its Executive Committee, both in the form of travel grants for some participants and in the form of moral support for the importance of education. Many other participants received support from their institutions or countries, and we are grateful to those who made sure that these individuals could attend and participate.

The work of the IAU Commission on Education and Development continues. Its breadth can be seen on the Commission's web page at http://www.astronomyeducation.org

The bulk of the work of formatting this book in L^AT_EX was done by Joseph B. Wilson at the University of Toronto. We thank him for his hard work and good judgment. We are grateful to Dr. Naomi Pasachoff for her careful reading of the manuscript and her corrections and suggestions. We also acknowledge a grant from the Natural Sciences and Engineering Research Council of Canada for partial support for the creation of this book.

Jay M. Pasachoff
John R. Percy

Introduction

The quantity and quality of the astronomy that is taught in our schools has a critical impact on the health of astronomy. It affects the recruitment and training of future astronomers. It affects the awareness, understanding, and appreciation of astronomy by the citizens who, as taxpayers and decision-makers, support our work. They form the society and culture within which we operate. In many countries, astronomy does not appear in the school curriculum at all; in other countries, it has a place in the curriculum, but the curriculum may be flawed, or teachers may have neither the training nor the resources to present it effectively. Much is known about effective teaching and learning of astronomy. Very little of this knowledge is implemented in schools and universities. Rather, teaching may be ineffective; it may sometimes intensify misconceptions, and may create an incorrect or negative impression of our subject.

Yet we live in a golden age of astronomy. In the last half-century, astronomers have explored dozens of planets and moons in our solar system, and astronauts have set foot on one moon – ours. Astronomers have discovered over a hundred planets around other stars. They have learned much about the life cycle of stars, including their bizarre end products – white dwarfs, neutron stars, and black holes. On a wider scale, they have mapped the universe of galaxies and, in the twenty-first century, they have determined the age, shape, and composition of the universe with unprecedented accuracy. We have begun to understand our cosmic roots: the origin of our universe, our galaxy, our star, and our planet, and of the atoms and molecules of life. We can speculate, with increasing confidence, about whether the same processes have given rise to life elsewhere in our universe.

Astronomers are not only obligated to share the nature and excitement of their discoveries with the public, but also most astronomers are keen to do so. Indeed, they are distressed if they find that the public is uninterested, uninformed, or misinformed. Fortunately, public interest tends to be high.

This conference brought together experienced and knowledgeable astronomers, astronomy educators and education researchers from around the world to (i) review what is known about effective teaching and learning of astronomy at the school level, and how it can be implemented; (ii) examine specific examples of successful (or unsuccessful) approaches to teaching; and (iii) provide guidance for improvement in the future. The emphasis was on identifying and implementing practices that are practical and widely applicable, taking account of contemporary education research, and the widespread interest in topics of astronomy, including current developments. The needs of the developing world were explicitly addressed. Large parts of the industrialized world could be considered undeveloped as far as astronomy education is concerned!

1

This conference focused on formal education in elementary and secondary school, since that is where the overwhelming majority of people may be exposed to astronomy. Pre-service and in-service teacher education were considered, as was introductory astronomy at the university level insofar as it affects general scientific literacy – especially among prospective teachers. The roles of planetariums, science centers, print and electronic media, professional and amateur astronomers in supporting school education were addressed, either explicitly or implicitly, as was the challenge of forming productive partnerships among astronomy educators, education researchers, teacher educators, and all the other individuals and organizations that have a role to play. The many forms of instructional technology (from robotic telescopes to the Internet) were highlighted, especially where they can be shown to promote effective teaching and learning.

One of the goals of this conference was to encourage more and better astronomy in schools around the world. A second goal, which will help to achieve the first goal, was to encourage and facilitate the development of teacher training in astronomy, and of resources and other materials for teachers. A third goal was to identify effective, efficient, culturally appropriate strategies for achieving these goals in each country. These goals are expressed in a Resolution that was presented to the 2003 IAU General Assembly by Commission 46. We are grateful to Magda Stavinschi (Romania) for starting the process which led to this Resolution. The spirit of her proposal is well expressed in the abstract of her conference paper "Why astronomy should be taught in schools – A resolution":

Perhaps more than other post-Communist countries, Romania underwent a dramatic change after 1983. One of the sharpest problems remains the change of mentality and, implicitly, education of youngsters. Astronomy has a particularly important role in this context, especially now. Unfortunately, it does not belong to curricula anymore. Its reintroduction, as well as a new system of scientific education in agreement with the knowledge and technology of the 21st century, is compulsory in both Romania and other European countries. This international campaign and new ways for implementing astronomy in education can be carried out after the adoption of an IAU resolution (see *Newsletter* 55, October 2001). The same problems, specific not only to Romania but to Europe, too, were tackled in special sessions on "Astronomy education in Europe" at JENAM (Joint European and National Astronomical Meeting) in 2003 and 2004.

See also:

http://www.konkoly.hu/jenam03/ for information on JENAM 2003 and
http://www.iaa.csic.es/jenam2004/ for JENAM 2004.
http://www.astro.ulg.ac.be/RPub/Colloques/JENAM/

The following Resolution was proposed to the 2003 IAU General Assembly by Commission 46, and passed unanimously by the National Representatives to the General Assembly. As mentioned, the Resolution was first suggested by Magda Stavinschi (Romania), and was further developed by the Organizing Committee of Commission 46, with important contributions by Johannes Andersen (Denmark):

Considering

   (1)   that scientific and mathematical literacy and a workforce trained in science and technology are essential to maintain a healthy population, a sustainable environment, and a prosperous economy in any country,

   (2)   that astronomy, when properly taught, nurtures rational, quantitative thinking and an understanding of the history and nature of science, as distinct from rote learning and pseudoscience,

(3) that astronomy has a proven record of attracting young people to an education in science and technology and, on that basis, to careers in space-related and other sciences as well as industry,

(4) that the cultural, historical, philosophical and aesthetic values of astronomy help to establish a better understanding between natural science and the arts and humanities,

(5) that, nevertheless, in many countries, astronomy is not present in the school curriculum and astronomy teachers are often not adequately trained or supported, but

(6) that many scientific and educational societies and government agencies have produced a variety of well-tested, freely-available educational resource material in astronomy at all levels of education.

[IAU] Recommends

(1) that national educational systems include astronomy as an integral part of the school curriculum at both the elementary (primary) and secondary level, either on its own or as part of another science course,

(2) that national educational systems and national teachers' unions assist elementary and secondary school teachers to obtain better access to existing and future training resources in astronomy in order to enhance effective teaching and learning in the natural sciences,

(3) that the National Representatives in the IAU and in its Commission 46 call the attention of their national educational systems to the resources provided by and in astronomy, and

(4) that members of the Union and all other astronomers contribute to the training of the new, scientifically literate generation by assisting local educators at all levels in conveying the excitement of astronomy and of science in general.

Implementation of this Resolution will be a major challenge. It will require effective linkages between astronomers and educators; the National Liaisons to Commission 46 can play an important role here. They can work through the "astronomical community" in each country: professional and amateur astronomers, and educators at all levels and in all settings. See www.astronomyeducation.org for a list of National Liaisons.

Successful learning of astronomy will occur if the curriculum is developed and delivered in accordance with the results of educational research, and with an appropriate degree of enthusiasm and excitement. Research is therefore as essential to successful learning of astronomy as it is to astronomers' quest for understanding of the universe.

Part I

Astronomy in the curriculum
around the world

Introduction

The place and nature of astronomy in the school curriculum vary greatly from one country to another, and even from place to place within a country. There are two main "systems" of education, which are usually called the "European" system and the "North American" system. These, and the place of astronomy in each, were eloquently described by Don Wentzel in his prologue to the proceedings of the 1988 IAU conference on astronomy education, held in Williams College, Williamstown, Massachusetts, USA.[1]

In the European system, there is usually a national curriculum. An often cited example is that in France, at 10 a.m. on the second Thursday in April, every student in a certain grade is learning the same thing from the same page in the textbook. Students are streamed, at an early age, into university, technical school, or the workplace. Astronomy tends to be taught to science students, by teachers who are well trained in science and science teaching.

In the North American system, the curriculum may be determined locally; astronomy is taught in a variety of places in the curriculum; and the teachers may therefore not be well trained in astronomy content or pedagogy. A recent requirement in the US's No Child Left Behind Act (a controversial law that played a role in the presidential campaign), that every teacher be certified in every subject that he or she is teaching, does not meet, for example, the reality of small schools in rural states in which a single teacher may teach some biology, some chemistry, and some physics. This requirement is, therefore, being relaxed. Only 7 per cent of the national total of school spending comes from the US federal government, but that "tail" is enough to "wag the dog."

In the North American system, a wide variety of entities – individual schools, cities, counties, or states or provinces, for example – decide on curriculum. The few states with major statewide adoptions of books or of an approved list of books, such as Texas and California, greatly influence national teaching by their choices through publishers' attempts to please them. No astronomy course is part of a widespread curriculum, though astronomy is taught as part of earth science to students of approximate age 14 on a widespread scale and on a much more minor scale at other periods. On high-school levels, an occasional astronomy course may be found, or some astronomy content may be taught – largely through the initiative of individual teachers – in physics or earth science courses. Though no recent statistics are available, a statistics expert at the American Institute of Physics informs us that, in 1986, only 3 per cent of all physics teachers also taught astronomy. (Their survey would

---

[1] See Pasachoff and Percy, 1990; the contents are newly available online in the Astrophysical Data System. To see any of the articles, go to: http://adswww.harvard.edu, choose "Search References," and then choose "Astronomy and Astrophysics." You can then ask for a table of contents and search by author name, title, or abstract key words.

not include teachers who taught only astronomy or astronomy plus non-physics subjects, but those numbers should be even lower.) In 1987, 11 per cent of physics teachers had taught a class on astronomy at some time, "one-third only once and more than three-quarters no more than five times during their teaching careers."

Guidelines to astronomy and other topics, such as that of the American Association for the Advancement of Sciences' *Project 2061* were written in the absence of substantial participation by astronomers and tend to shortchange our science (Pasachoff, 1996, 1998). The project has great influence on curriculum. See *Benchmarks for Science Literacy* (1994) and *Designs for Science Literacy* (2001), available from Oxford University Press (www.oup-usa.org). The former "specifies what student progress is reasonable to have been made by the end of Grades 2, 5, 8 and 12," while the latter gives general guidance on making curricula. See also the general Project 2061 website at http://www.project2061.org. Actual curricula, and the textbooks that are used to teach them, depend greatly on a range of conflicting requirements from many of the 50 US states (Pasachoff, 2002; and Pasachoff's "Textbooks for K–12 astronomy," Chapter 8 in this volume).

In practice, there is a great diversity in the place of astronomy in the school curriculum worldwide. One source of information is the Triennial Reports that are submitted by the National Liaisons to IAU Commission 46 on Education and Development. They can be found on the Commission's website: http://www.astronomyeducation.org, which is equivalent to http://physics.open.ac.uk/IAU46. Another source is the proceedings of the previous IAU colloquia on astronomy education, held in 1988 in Williamstown, Massachusetts (Pasachoff and Percy, 1990), and in London, UK, in 1996 (Gouguenheim, McNally, and Percy, 1998), or of the 1994 IAU session on the subject (Percy, 1994). The Astronomical Society of the Pacific (www.astrosociety.org), based in California, often provides curriculum-related content at its annual meetings, though the emphasis is on the North American system. Many ideas appeared in an ASP symposium on *Astronomy Education: Current Developments, Future Coordination*, held in College Park, Maryland, in 1995 (Percy, 1996). An ASP symposium on teaching astronomy, entitled *Cosmos in the Classroom 2004*, took place at Tufts University in Medford, Massachusetts, in July 2004. Though it dealt only with levels from advanced high school upwards, it stressed hands-on involvement, a technique that lends itself to lower grades as well, which are the focus of this book.

In Chapter 1, John Percy lists the many reasons why astronomy is useful and should be included in the school curriculum, somewhere. The place of astronomy in the curriculum may, therefore, depend on why the curriculum developers think that astronomy is important (or not). If they feel that the practical effects of the cosmos on the Earth are most important, then astronomy may appear as part of a geography course. If they feel that its role as a basic science, or a part of physics, is most important, then it may appear as part of a physics course. In Chapter 2, Rosa M. Ros describes the important role that astronomy can play within a mathematics course, as a tool, as an application, and as an inspiration for students.

If astronomy is felt to be an important discipline in its own right, then it may appear as a stand-alone course, or a stand-alone section of a general course such as "science." Astronomy is rich in interdisciplinary and therefore cross-curricular connections, and these are highly valued in modern curriculum design. These connections are much appreciated, especially by elementary-schoolteachers, who tend to be generalists in their background, and teach all subjects in the curriculum. A major question is how to get a wide range of elementary-schoolteachers – or, at least, students at faculties of education – familiar enough with astronomy to feel at ease in teaching it. This issue is addressed later in this volume.

Assuming that astronomy does appear in the curriculum, there is the question of which topics should be included. Again, this will be influenced by the beliefs of the curriculum designers, and the teachers: which of the "uses" of astronomy are most important? If it is the cultural and philosophical connections, then there should be a good dose of history in the course. If it is the physical and mathematical connections, then there should be measurements to make and problems to solve. If it is the aesthetic and environmental connections, then there should be observation of the real sky and/or contemplation of astronomical images. It then becomes easy to argue that all topics should be included. That may be possible for very wise teachers and students, but it usually leads to "curriculum overload."

One approach to curriculum overload is to teach basic astronomical concepts and terminology, so that students can read about and understand the details, including current developments. "Basic" concepts should therefore include concepts such as galaxies and black holes – not just moon phases and seasons. Another approach is to choose, carefully, a small number of topics and activities that touch upon all the expectations of the curriculum – knowledge, skills, applications, and attitudes. These topics would include both "classical" and "modern" ones. The skills would include generic skills, such as understanding the scientific process, that could be applied to other topics and activities. Those topics and activities that were covered would then be taught in sufficient depth so that students learned them successfully. There is no one correct answer for what aspects of astronomy the curriculum should contain. We can only hope that astronomy is there in some form and in some guise.

References

Gouguenheim, L., McNally, D., and Percy, John R., eds. 1998, *New Trends in Astronomy Teaching*, IAU Colloquium 162, Cambridge: Cambridge University Press.

Pasachoff, Jay M. 1996, "Remarks on Project 2061," in John R. Percy, ed., *Astronomy Education: Current Developments, Future Coordination*, ASP Conference Series, **89**, San Francisco: Astronomical Society of the Pacific, 30–2.

Pasachoff, Jay M. 1998, "Astronomy and the new National Science Education Standards," *Forum on Education of the American Physical Society Newsletter*, Spring 1998, 13–14; also printed in the *Newsletter of the American Astronomical Society*, June 1996, **80**, 8–9, as Jay M. Pasachoff with Jason Lorentz, "Astronomy and the new National Science Education Standards: some disturbing news and an opportunity."

Pasachoff, Jay M. 2002, "Astronomy textbooks," Proceedings of "Communicating Astronomy," held at Astrophysics Institute of the Canary Islands, http://www.iac.es/proyect/commast/; and tmj@ll.iac.es.

Pasachoff, Jay M. and Percy, John R., eds. 1990, *The Teaching of Astronomy*, Cambridge: Cambridge University Press.

Percy, John R. 1994, *Current Developments in Astronomy Education*, based on Joint Discussion No. 4 of the IAU General Assembly, The Hague, Netherlands.

Percy, John R., ed. 1996, *Astronomy Education: Current Developments, Future Coordination*, ASP Conference Series, **89**, San Francisco: Astronomical Society of the Pacific.

I

Why astronomy is useful and should be included in the school curriculum

John R. Percy

University of Toronto, Mississauga, ON, Canada, L5L 1C6

Abstract: Why is astronomy useful? Why should it be part of the school curriculum? This paper lists about 20 reasons: cultural, historical, and philosophical reasons; practical, technological, and scientific reasons; environmental, aesthetic, and emotional reasons; and pedagogical reasons. Astronomy can attract young people to science and technology; it can promote public awareness, understanding, and appreciation of science; it can be done as an accessible, inexpensive hobby: "the stars belong to everyone." This paper then connects the reasons to the expectations of the modern school curriculum, including knowledge, skills, applications, and attitudes.

One of the goals of the conference upon which this book is based is to encourage more and better astronomy in schools around the world. A second goal, which will help to achieve the first goal, is to encourage and facilitate the development of teacher training in astronomy, and of resources and other materials for teachers. A third goal is to identify effective, efficient, culturally appropriate strategies for achieving these goals in each country. These goals are expressed in the Resolution which was presented to the 2003 IAU General Assembly by Commission 46. I am grateful to Magda Stavinschi, of Romania, for starting the process which led to this Resolution (see the Introduction to this volume). Implementation of these goals will require effective linkages between astronomers and educators; the National Liaisons to IAU Commission 46 can play an important role here. They can work through the "astronomical community" in each country, as defined by Percy (1999).

We must first know what the goals of the school curriculum in each country or region are. In my country of Canada, education is a provincial responsibility, though there has been some national co-ordination in the area of science education. The stated goals of the grade 1–8 (age 6–13) school science curriculum in Ontario are:

to understand the basic concepts of science and technology; to develop the skills, strategies, and habits of mind required for scientific inquiry and technological design; and to relate scientific and technological knowledge to each other, and to the world outside the school.

The purposes of these goals are "to enable the students to be productive members of society ... and to develop attitudes that will motivate them to use their knowledge and skills in a responsible manner." At the grade 9–10 (age 14–15) level, the goals are similar, except for one addition: "to relate science ... to the environment." The overall aim is "to ensure scientific literacy for every secondary school graduate" (since some graduates may not study science beyond the grade 10 level). At the grade 11–12 (age 16–17) level, the goals and the overall aim remain the same, but the courses and their content are now tailored to the students' future paths – to university, to colleges of applied arts and technology, or to the workplace.

The science curriculum thus has four elements: science, technology, society, and environment. And it has four sets of expectations: knowledge, skills, applications, and attitudes. The last of these includes ethical issues. I would like to think that it also includes an appreciation of the cultural, aesthetic, and emotional aspects of science – all of which are relevant to astronomy.

I should stress, however, that I have been discussing "the intended curriculum." This is not the same as "the implemented curriculum." In the intended curriculum, astronomy is allocated one-quarter of the science curriculum in grades 6 and 9. In practice, teachers may leave astronomy to the last week or two of the year, or not cover it at all. And there is also a difference between what is taught, and what is learned. Education researchers have showed convincingly that despite (or sometimes because of) teaching, students actually hold deeply rooted misconceptions about astronomical topics. They believe, for instance, that the seasonal changes in temperature result from the changing distance of the Earth from the sun.

It would be interesting to start by asking: "why is astronomy *not* included in the curriculum?" Here are some possible reasons; I thank my colleagues in IAU Commission 46 for their comments on these:

- astronomy is perceived to be irrelevant to practical concerns such as health, nutrition, agriculture, environment, engineering, and the economy in general; this is particularly true in developing countries;
- most schoolteachers have little or no knowledge of astronomy, or astronomy teaching; in fact, they may have the same deeply rooted misconceptions as their students;
- astronomy is perceived as requiring night-time activities ("the stars come out at night, the students don't"), and expensive and complex equipment such as telescopes;
- astronomy is perceived as being solely "Western" by some non-Western cultures;
- there may be conflict – real or perceived – between astronomy and personal beliefs such as religion, culture, and pseudoscience; in fact, astronomy is sometimes viewed as being as speculative as these fields are;
- many of the available resources are designed for affluent schools in affluent countries, or for different latitudes, longitudes, and languages;
- astronomy may be seen as allied with high technology, with all its real and perceived dangers.

Many of these reasons are based on a lack of an astronomical "tradition" in a country or region. This is one more reason for all members of "the astronomical community" to speak and work together in promoting astronomy.

Now we can address the main topic of this presentation: the reasons why astronomy is useful, and should be part of the school curriculum – in science, or some other place. These reasons can be divided, broadly, into several groups:

- Astronomy is deeply rooted in almost every culture, as a result of its practical applications, and its philosophical implications.
- Among the scientific revolutions of history, astronomy stands out. In the recent lists of "the hundred most influential people of the millennium," a handful of astronomers were always included.
- Astronomy has obvious practical applications to timekeeping; calendars; daily, seasonal, and long-term changes in weather; navigation; the effect of solar radiation, tides, and impacts of asteroids and comets with the Earth.

- Astronomy is a forefront science that has advanced the physical sciences in general by providing the ultimate physical laboratory – the universe – in which scientists encounter environments far more extreme than anything on Earth. It has advanced the geological sciences by providing examples of planets and moons in a variety of environments, with a variety of properties.
- Astronomical calculations have spurred the development of branches of mathematics such as trigonometry, logarithms, and calculus; now they drive the development of computers: astronomers use a large fraction of all the supercomputer time in the world.
- Astronomy has led to other technological advances, such as low-noise radio receivers, detectors ranging from photographic emulsions to electronic cameras, and image-processing techniques now used routinely in medicine, remote sensing, etc. Its knowledge is essential as humankind continues to explore outer space.
- Astronomy, by its nature, requires observations from different latitudes and longitudes, and thus fosters international co-operation. It also requires observations over many years, decades, and centuries, thus linking generations and cultures of different times.
- Astronomy reveals our cosmic roots, and our place in time and space. It deals with the origins of the universe, galaxies, stars, planets, and the atoms and molecules of life – perhaps even life itself. It addresses one of the most fundamental questions of all – are we alone in the universe?
- Astronomy promotes environmental awareness, through images taken of our fragile planet from space, and through the realization that we *may* be alone in the universe.
- Astronomy reveals a universe that is vast, varied, and beautiful – the beauty of the night sky, the spectacle of an eclipse, the excitement of a black hole. Astronomy thus illustrates the fact that science has cultural as well as economic value. It has inspired artists and poets through the ages.
- Astronomy harnesses curiosity, imagination, and a sense of shared exploration and discovery (I think Ontario science teacher Doug Cunningham was the first to put this so eloquently).
- Astronomy provides an example of an alternative approach to "the scientific method" – observation, simulation, and theory, in contrast to the usual experiment and theory approach.
- Astronomy, if properly taught, can promote rational thinking, and an understanding of the nature of science, through examples drawn from the history of science, and from present issues such as pseudoscience;
- Astronomy, in the classroom, can be used to illustrate many concepts of physics, such as gravitation, light, and spectra.
- Astronomy, by introducing students to the size and age of objects in the universe, gives them experience in thinking more abstractly about scales of time, distance, and size.
- Astronomy is the ultimate interdisciplinary subject, and "integrative approach" and "cross-curricular connections" are increasingly important concepts in modern school curriculum development.
- Astronomy attracts young people to science and technology, and hence to careers in these fields.
- Astronomy can promote and increase public awareness, understanding, and appreciation of science and technology, among people of all ages.
- Astronomy is an enjoyable, inexpensive hobby for millions of people.

References

Percy, J. R. 1999, *Teaching Astronomy in the Asian-Pacific Region*. Bulletin No. 15, 77–80.

Comments

Jay Pasachoff: I am sorry to see how sporadic the inclusion of astronomical topics is in the Canadian curriculum. It is much less evenly distributed than in the United States' curriculum that I will discuss [Chapter 8]. All the topics covered in the early years, as shown in your list, are positional or factual – "what" rather than "why." Yet when I meet a third-grade class, the students eagerly ask about black holes and other topics of contemporary interest. I think that the elementary and intermediate students should be taught about interesting modern topics, so as not to stifle their interest. Good textbooks and teacher training are important.

John Percy: Astronomy may be evenly distributed in the USA in the "intended curriculum," but not in the "implemented curriculum." And we have good textbooks in Canada, also; I have worked on some of them.

Ragbir Bhathal: One of the problems is to get the Department of Education to provide funds for teacher education in astronomy during the school term. We also need actual hands-on detailed instructions for astronomy projects, i.e., photometric studies of variable stars, etc.

John Hearnshaw: Could you comment on two points: (1) Is astronomy economically beneficial to a country? and (2) Should we push astronomy in the high school curriculum without first developing students' understanding of maths and physics to a high level?

John Percy: The main economic benefits of astronomy are probably indirect – through its effect on students' awareness and appreciation of science and technology. But astronomy can promote high-tech industry, in computing and communication and aerospace, for instance.

Julieta Fierro: I believe one should not only train teachers but also the parents by way of outreach programs.

John Percy: The Astronomical Society of the Pacific has recently developed *Family ASTRO* for families. And the Canadian Astronomical Society has developed the online Canadian Junior Astronomer Program for young people both in and outside of school.

Vikram Ravi: The main factor that decides the appeal and interest of astronomy to high school age students is the background and the level of interest of the teachers in astronomy. Most teachers in my experience have done undergraduate science, but have majored in chemistry and biology (even the physics teachers!). Surveys in North America have shown this as well.

2

Astronomy and mathematics education

Rosa M. Ros

Technical University of Catalonia, Vilanova (Barcelona), E-08800, Spain

Abstract: There are many European countries in which astronomy does not appear as a specific course in secondary school. In such cases, astronomy can be introduced through other subjects. This paper concerns astronomy in mathematics classes. Teaching astronomy through mathematics would result in more exposure than through physics, for instance, because mathematics is more prevalent in the curriculum. Generally, it is not easy to motivate students in mathematics, but they are motivated to find out more about the universe, and about astronomy current events which appear in the media. This provides an excellent introduction to several mathematics topics. Specific connections include: angles and spherical coordinates to star trails; logarithms to visual magnitudes; plane trigonometry to orbital motion; spherical trigonometry to the obliquity of the ecliptic; and conic curves to sundials at various latitudes. These practical, applied connections make mathematics courses more attractive to students.

2.1 Astronomy in mathematics in schools

Astronomy has an interdisciplinary aspect that is potentially very positive. It is good to combine astronomy with other topics and to introduce astronomy in general projects in school in order to integrate several courses, for instance physics, mathematics, geography, biology or history. However, these kinds of project are sporadic. Of course, it is a good idea to promote them, but it is not possible to organize the astronomical education of young people with only this kind of project. It can be positive for astronomy to appear as a small part of other subjects. If astronomy appears in several subjects, this communicates to the students astronomy's interdisciplinary nature, which is very positive. The fact that astronomy is present in so many different areas of human knowledge is proof in itself of its interest. As a consequence, astronomy should appear always and everywhere in school.

However, it can also be negative for astronomy to appear in a lot of subjects. The problem is simple and real. If teachers are non-specialists, perhaps they favor other parts of the syllabus. Teachers dislike teaching aspects that they do not know very well and try to avoid them. In this case the situation has reversed. Astronomy is never and nowhere. In some cases this second option is more common than we would like. It is not unusual for some teachers to decide to cover the astronomy part of the curriculum by going to the planetarium with their students. In this case, in one morning the students receive all the astronomical content compressed. This is pointless. They can assimilate practically nothing. Of course the planetarium is an excellent resource for the astronomy class, but it can never be a substitute for teacher input in the classroom.

To educate society in astronomy it is necessary to teach astronomy at primary and secondary school. Students may then decide to study astronomy at university. It is hard to imagine students deciding to study a degree in a topic that they do not know. The best way to teach astronomy in school is to introduce it as a course in its own right. The viability of this solution depends on the national curriculum and the plans of the education ministry in each country. In general, the majority of countries do not have astronomy as a specific course and it is not easy to change this. Therefore, if this option is not possible, we should adapt to the real situation, that is to say include astronomy as a part of other courses. Currently there is astronomical content in physics, natural science, and social science, depending on the country. This paper aims to propose a new point of view that could work to help bring about the integration of astronomy into the curriculum.

Commonly, physics teachers include some astronomy in their classes. In some cases it is possible that other science teachers (biologists or geologists) or non-science teachers (geographers or historians) also teach astronomy. Of course, in order to promote astronomy, we have to promote all these possibilities; however, the focus of this paper is the introduction of astronomy content within mathematics. This idea has several advantages and very few disadvantages. If mathematics teachers try to introduce astronomy into their classes, the situation will be better for everyone, because astronomy will be promoted, and mathematics enriched.

Advantages for mathematics

- **Astronomy will make mathematics more attractive.** Mathematics is not very attractive to most pupils; astronomy is more motivating. The schoolteacher knows very well that the students are not very interested in mathematics. *Students are not interested in discovering what an ellipse is, but they are really concerned about the possibility of the moon crashing into the Earth.* The objective can be to present the mathematical content connected with astronomical objects or situations. Students are often very keen to ask astronomy teachers questions. Everyone has questions about the universe. It is very unusual for students to have a question for instance about "trigonometry," but it is surprising the number of people who would like to know if the Earth could really be destroyed by a comet crashing into it. Maybe this is because of the movies, but this is unlikely. These kinds of questions are as old as humanity. People find astronomy attractive. It is one of the few science subjects that enjoys great coverage in the media. This fact is very significant.

- **Astronomy connects mathematics with the real world.** Schools do not have "mathematics laboratories." Mathematics is a very theoretical subject for students and in general seems unconnected with the real world. Lay people think that mathematics is necessary to count, but practically nothing more. Anything more than basic arithmetical rules seems beyond them. In reality, mathematics is much more. It is present in the nucleus of all science. If the students become scientists, they will need to use mathematics. So, they need to study mathematics, but they want to know why they need it, and what content they need. What they really want to know is what contact mathematics has with reality. This is not an easy question to answer. It is necessary to put mathematics back into contact with the real world. It is necessary to take mathematics back to the lab, but the schools do not have a mathematics lab.

  Astronomy offers mathematics teachers a laboratory: the school grounds. If it is not cloudy, students can observe the sky, take measurements, and analyze the results obtained with their mathematical knowledge (taking them beyond arithmetic). It is also possible to introduce scientific methodology and to study the margin of error.

Students are not very interested in calculating the third side of a triangle from the other sides and angles, but they will be interested in finding the distance between the Earth and the moon after making their own observations. Maybe the results are not very accurate, but the experience will be very motivating, and they never will forget it (see an example in the last section in this paper). This is wonderful for students. They can get enough information themselves in order to calculate (approximately) the distances Earth–moon and Earth–sun as well as their diameters. This is a simple exercise that it is possible to do at secondary school using astronomy and some simple mathematical content that they have in their curricula.

Advantages for astronomy

- **Astronomy would be present from the beginning to the end of school.** Mathematics appears from the first course of primary school until the end of secondary school. Students study mathematics for many years, providing an enormous opportunity to introduce many different astronomical ideas, whereas, if astronomy appears only in physics courses, it is introduced very late in the curriculum. It is, therefore, better to also introduce astronomy in mathematics courses. Of course, there is some content in humanities courses, but in these cases the teacher is probably not an astronomy enthusiast. This is too much "science" for them, and they prefer to keep astronomy content to a minimum or cut it altogether. If the teacher is a non-specialist it's very possible that astronomy never appears. *8–9-year-old students are very curious. They ask a lot about their surroundings. Why does the moon look like a croissant? And, why sometimes not? Why can we see the moon in the morning? (In children's stories the moon only appears at night.)* We are currently missing an opportunity to harness the curiosity of these younger students as these questions aren't answered until they study physics in secondary school. This curiosity was the origin of the explorers' trips and the researchers' work. It is necessary to promote this curiosity. Especially younger students ask a lot of questions about the universe and what surrounds our planet, and the teacher most connected with the scientific field at primary school is often the mathematics teacher. It is not difficult to promote astronomy by means of these teachers with good training.
- **Astronomy is presented as a quantifiable subject as opposed to a pseudoscience.** *A lot of students think that astronomy and astrology are synonymous. It is not only young people. If you request an astronomy book in a book shop, it is really possible that they will direct you to the astrology section.*

  It is good for astronomy to appear inside math classes because students are left with no doubt that astronomy is a serious and rigorous subject. The teacher can present astronomy as a subject that requires observation, the taking of measurements, the analysis of data, the making of deductions, the study and verification of results and the analysis of errors. This is very different from astrology content, which is not possible to verify and does not offer a scientific method.

Disadvantages for mathematics and astronomy

- **If the astronomical content or the mathematical application is too difficult, the effect could be "counter-productive" for both subjects.** We have to be prudent and choose the examples very carefully. It is not a good idea to have to explain a lot of astronomical content in order to make an example. This can be boring for the students and in this case the results are negative. It is the same idea about the mathematical information that they

Fig. 2.1. A teacher explains Kepler's Laws at a mathematics conference (Coruna 7).

need. It is not a good idea to explain a new mathematical idea in order to use it in an astronomical activity. The objective is to use the information that they already know in order to achieve new information relating to the universe.

Conclusions

It is very important to offer mathematics teachers a good selection of topics and activities connecting astronomy and mathematics. If they have good materials they will use them. Of course, it is important to select very carefully a collection of astronomical items that are interesting from a mathematical point of view, according to the curriculum and related to the real world, in order to attract the students. We have to make sure students and teachers are satisfied if we want to achieve success.

This option is not incompatible with teaching astronomy in physics courses. The interests of physics and mathematics teachers are not the same. This offers us two different but equally valuable points of view. Several examples of suitable materials for mathematics teachers are:

- angles and star trails (the same angle for the same exposure time in a photograph of the sky);
- logarithms and visual magnitudes (students are surprised when they discover that their eyes know logarithms, and perhaps they failed the logarithms exam in the mathematics course);
- plane trigonometry related positions (related orbital movements);
- spherical trigonometry and ecliptic obliquity (relationships on ecliptic obliquity and latitude);

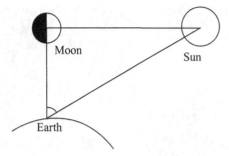

Fig. 2.2. Relative position of the moon's quarter.

- conic curves related to sundials (by finding conic curves in the street);
- interpolation and variable stars (by finding the magnitudes of variable stars);
- graph of functions and "light curves" (for variable stars).

The next section outlines activities that may be used to introduce astronomy in schools. The bibliography also includes some examples (only in English, French and Spanish).

2.2   An example: Aristarchus' and Eratosthenes' experiments revisited

Aristarchus (310 BCE–230 BCE) deduced some proportions between the distances and the radii of the Earth–moon–sun system. Aristarchus calculated the sun's radius, the moon's radius, and the distances Earth–sun and Earth–moon in relation to the Earth's radius. Some years afterwards, Eratosthenes (280 BCE–192 BCE) determined the radius of our planet, and it was possible to calculate all the distances and radii of the Earth–moon–sun system. This proposed activity is to repeat both experiments with students. The idea is to repeat the mathematical process designed by Aristarchus and Eratosthenes and, if possible, to repeat the observations.

2.2.1   *Aristarchus' experiment redone*

- **Relationship between the Earth–moon and Earth–sun distances.** Aristarchus determined that the angle between the sun and the moon, as seen from Earth, when the moon was in the quarter phase position, was 87° (Fig. 2.2). We know that he made a mistake, possibly because it was very difficult for him to determine the instant of the quarter phase. In fact this angle is 89°51′, but the process used by Aristarchus is correct. In Fig. 2.2, if we use the definition of the sine, it can be deduced that $\sin 9' = \frac{ES}{EM}$, where ES is the distance from the Earth to the sun, and EM is the distance from the Earth to the moon. Then approximately,

$$ES = 400\,EM$$

(although Aristarchus deduced ES = 19 EM).
- **Relationship between the radius of the moon and the radius of the sun.** The relationship between the diameters of the moon and the sun must be similar to the previous formula, because from the Earth we observe both diameters equal to 0.5°. Therefore, the radius also verifies

$$R_S = 400\,R_M$$

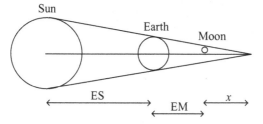

Fig. 2.3. Shadow cone and relative positions of the Earth–moon–sun system.

- **Relationship between the Earth's distance from the moon or the sun and the radius of the moon or the sun.** Given that the moon's diameter observed is 0.5°, with 720 times this diameter it is possible to cover the circular path of the moon around the Earth. The length of this path is 2π times the distance Earth–moon, that is to say, $2R_M\,720 = 2\pi$ EM. Isolating,

$$\text{EM} = \frac{720}{2\pi}R_M$$

and, by similar reasoning,

$$\text{ES} = \frac{720}{2\pi}R_S$$

- **Relationship between the Earth's distances from the moon and sun, the moon's radius, the sun's radius and the radius of the Earth.** During a moon's eclipse, Aristarchus observed that the time necessary for the moon to cross the shadow cone of the Earth was double the time necessary for the moon's surface to be covered (Fig. 2.3). Therefore, he deduced that the shadow of the Earth's diameter was double the moon's diameter, that is to say, the relation of both diameters or radii was 2 : 1. Really, we know that this value is 2.6 : 1. Then in Fig. 2.4 we deduce the following relationship:

$$\frac{x}{2.6R_M} = \frac{x + \text{EM}}{R_E} = \frac{x + \text{EM} + \text{ES}}{R_S}$$

where x is an auxiliary variable. Introducing in this expression the relationships ES = 400EM and $R_S = 400R_M$, it is possible to eliminate x. Simplifying gives

$$R_M = \frac{401}{1440}R_E$$

which offers the opportunity to express all the dimensions in terms of the Earth's radius, so

$$R_S = \frac{2014}{18}R_E$$

$$\text{ES} = \frac{81640}{\pi}R_E$$

$$\text{EM} = \frac{401}{2\pi}R_E$$

We only need to calculate the Earth's radius to obtain all the distances and radii of the Earth–moon–sun system.

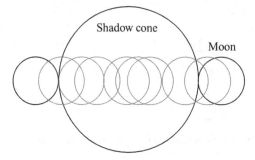

Fig. 2.4. Measuring the diameter of the shadow cone.

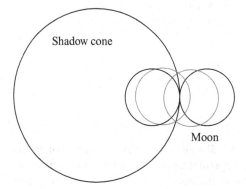

Fig. 2.5. Measuring the diameter of the moon.

- **Observations with students.** It is a good idea to repeat the observations made by Aristarchus with students. In particular, the students first have to calculate the angle from the Earth, between the quarter moon and the sun. For this measurement it is only necessary to use a theodolite and to find out the exact instant of the quarter phase. So we verify if this angle is 87° or 89°51′ (really it is very difficult). Secondly, during an eclipse of the moon, using a chronometer, it is possible to calculate the relation of times between "the first and the last contact of the moon with the Earth's shadow," that is to say, measuring the diameter of the shadow cone of the Earth (Fig. 2.4), and "the time necessary to cover the moon surface," that is to say, the measure of the moon's diameter (Fig. 2.5). Finally it is possible to verify if the relationship between both times is 2 : 1 or 2.6 : 1.

The most important objective of this activity is not the result obtained. The most important thing is that it suggests to the students that, if they use their knowledge and intelligence, they can obtain interesting results with few resources. In this case, the ingenuity of Aristarchus was important in obtaining some concepts of the size of the Earth–moon–sun system. It is a good idea to measure the Earth's radius with students according to the process used by Eratosthenes. Although Eratosthenes' experiment is very well known, we present here a short version in order to complete the previous Aristarchus experiment.

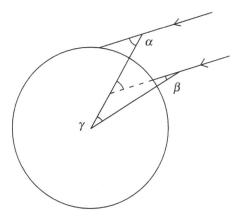

Fig. 2.6. Sticks' situation and angles in Eratosthenes' experiment.

2.2.2 Eratosthenes' experiment redone

We consider two sticks introduced perpendicularly into the soil, in two different cities on the Earth's surface on the same meridian. The sticks must be pointed towards the Earth's center. Normally it is better to use a plumb where we mark a point to measure lengths. We need to measure the length on the plumb from the soil to this mark, and the length on its shadow from the base of the plumb to the mark's shadow. We can consider that the solar rays are parallel. The sun's rays produce two shadows, one for each plumb. We measure the lengths of the plumb and its shadow. Using the definition of tangent we can obtain the angles α and β (Fig. 2.6). The central angle γ can be calculated because the sum of the three angles of a triangle is equal to π radians. Then, in the bottom triangle in Fig. 2.6, $\pi = (\pi - \alpha) + \beta + \gamma$ and, simplifying,

$$\gamma = \alpha - \beta$$

where α and β can be obtained from measuring the plumb and its shadow. Finally by proportionality between the angle γ and the length of its arc d (determined by the distance on the meridian between the two cities), and 2π radians of the meridian circle and its length $2\pi R_E$, that is to say $\frac{\gamma}{d} = \frac{360}{2\pi R_E}$, then

$$R_E = \frac{d}{\gamma}$$

can be deduced, where γ is obtained from the observation and d is the distance in kilometres between both cities. We can find d from a good map. The objective of this activity is not to obtain a precise result. We would only like the students to discover that by thinking and using all the resources that they can imagine, they can obtain surprising results.

Further reading

Broman, L., Estalella, R., and Ros, R. M. 1997, *Experimentos de Astronomía*, Alhambra, Mexico.
Broman, L., Estalella, R., and Ros, R. M. 1998, *Experimentos de Astronomía*, Alhambra, Madrid.
Camino, N. and Ros, R. M. 1997, "Por dónde sale el Sol?" *Educación en Ciencias, I*, Buenos Aires, Argentina, **3**, 11–17.

Esteban, E. and Ros, R. M. 1997, "Un simulateur de cadran solaire," *Les Cahiers Clairaut*, Orsay, Francia, **80**, 46–57.

Esteban, E. and Ros, R. M. 1996a, *Un ejemplo de cónicas (I)*, Universo, Barcelona, **16**, 88–91.

Esteban, E. and Ros, R. M. 1996b, *Un ejemplo de cónicas (II)*, Universo, Barcelona, **17**, 62–5.

Martins, F. and Ros, R. M. 2002, "Building simple sundials," *Proceedings of the 6th EAAE Summer School*, Barcelona, 59–82.

Ros, R. M. 1989, "Enseñar Astronomía con la ayuda de una cámara fotográfica," *Actas Congreso de Didáctica de Física, Microelectrónica y Microordenadores*, Madrid: UNED, 15–17.

Ros, R. M. 1990, "Prácticas de Astronomía a través de la fotografía," *Astronomía: Actualidad e Historia*, Zaragoza University, 181.

Ros, R. M. 1991, "Une étude d'étoiles variables," *Les Cahiers Clairaut*, **56**, 10–14.

Ros, R. M. 1992a, "Several astronomical ideas from Colon," *Teaching of Astronomy in Asian-Pacific Region*, Mitaka, Tokyo, **5**, 12–17.

Ros, R. M. 1992b, "Nociones de Fotografía. Aplicación a la Rotación Terrestre," *Astronomia: Aspectos Científicos y Culturales*, Zaragoza University, 117–32.

Ros, R. M. 1993, "Matemáticas al Sol," *Enseñanza de las Ciencias*, Bellaterra, IV, 347–348.

Ros, R. M. 1994a, "Pocket sundials," *Teaching of Astronomy in Asian-Pacific Region*, Mitaka, Tokyo, **9**, 1–20.

Ros, R. M. 1994b, "La Astronomía en el Diario de a Bordo de Cristóbal Colón," *Proceedings of International Symposium "Time and Astronomy at the Meeting of Two Worlds,"* Warszawa: Studia i Materialy, **10**, 453–459.

Ros, R. M. 1995a, "Logarithmic relation of magnitudes," *Teaching of Astronomy in Asian-Pacific Region*, Mitaka, Tokyo, **10**, 12.

Ros, R. M. 1995b, *Astronomía, Libros y Aventuras*, Barcelona: ICE, UPC.

Ros, R. M. 1996a, "Astronomical photography and Galileo's observations," *Manuskripte zur Chemiegeschichte*, Halle: TeaComNews, **7**, 7–38.

Ros, R. M. 1996b, "Matemática aplicada y relaciones de proporcionalidad," *Investigación e Innovación en Educación Matemática*, Santafé de Bogotá: EMA, **2**, 125–39.

Ros, R. M. 1996c, "Determinación de las magnitudes del sistema Tierra-Luna-Sol," *Astronomía y Astronáutica*, Universo, Barcelona, **9**, 74–77.

Ros, R. M. 1997, "Teaching youngsters about the moon's surface with limited facilities," *International Amateur-Professional Photoelectric Photometry*, **69**, 9–12.

Ros, R. M. 1998a, "Kepler's Laws and using photography to obtain numerical results," in *Mysterium Cosmographicum 1596–1996*, Prague, 288–96.

Ros, R. M. 1998b, *Estudio de la superfície lunar*, Universo, Barcelona, **39**, 62–7.

Ros, R. M. 1998c, "Teaching astronomy at secondary school level in Europe," in L. Gouguenheim, D. McNally, and J. R. Percy, eds., *New Trends in Astronomy Teaching*, IAU Colloquium 162, Cambridge: Cambridge University Press, 286–91.

Ros, R. M. 1999a, "Teaching several themes relating to inner and outer planets," *European Journal of Physics*, **20**, 331–41.

Ros, R. M. 1999b, *Investigando la superfície lunar*, Correo del Maestro, Mexico D.F., **34**, 24–37.

Ros, R. M. 1999c, "Actividades de geometría y matemática aplicada," *Investigación e Innovación en Educación Matemática*, Santafé de Bogotá, **5**, 51–67.

Ros, R. M. 2002a, "Old Egyptian constellations and Precession," in *Proceedings of the 6th International Conference Teaching Astronomy*, Barcelona, 237–41.

Ros, R. M. 2002b, "Sunrise and sunset positions change every day," in *Proceedings of the 6th EAAE SS Zaragoza University*, Barcelona, 177–88.

Ros, R. M., and Delpeix, J. 1990, "Órbitas y colas de cometas," *Tribuna de Astronomía y Universo*, Madrid, **6**, 70–8.

Ros, R. M. and Delpeix, J. 2001, "Estudio de los Cometas Hyakutake y Hale-Bopp," *Tribuna de Astronomía y Universo*, Madrid, **21**, 32–9.

Ros, R. M. and Estalella, R. 1987, "Aproximación didáctica al uso del telescopio: Satélites Galileanos de Júpiter," *La Escuela en Acción*, Madrid, **10477**, III, 20–4.

Ros, R. M. and Fabregat, J. 1994, "Some kinds of conics shadows," *Teaching of Astronomy in Asian-Pacific Region*, Mitaka, Tokyo, **8**, 19–27.

Ros, R. M. and Llinàs, M. J. 1989, "Determination of the obliquity of the ecliptic," *IAU Commission 46 Newsletter*, Liège, **28**, 11–13.

Ros, R. M. and Llinàs, M. J. 1991, "El eje del mundo está torcido," *Tribuna de Astronomía*, Madrid, **64**, 52–5.

Ros, R. M. and Llinàs, M. J. 1992a, "An astronomical activity without telescope: sunsets," in *Proceedings of "Teaching About Reference Frames: from Copernicus to Einstein,"* 423–7.

Ros, R. M. and Llinàs, M. J. 1992b, "An astronomical activity with telescope: Jupiter," in *Proceedings of "Teaching About Reference Frames: from Copernicus to Einstein,"* Torún, 428–32.

Ros, R. M. and Llinàs, M. J. 1992c, "La cámara fotografica, un instrumento astronómico," *Tribuna de Astronomía*, Madrid, **79**, 32–5.

Ros, R. M. and Llinàs, M. J. 1992d, "Saturno tras el objetivo de una cámara," *Tribuna de Astronomía*, Madrid, **84**, 18–20.

Ros, R. M. and Llinàs, M. J. 1992e, "Teaching Astronomy through photography," in *Proceedings of Teaching Astronomy; IVth International Conference*, Barcelona: UPC, 45–46.

Ros, R. M. and Viñuales, E. 1992a, "Algol, changing spirit," *IAU Commission 46 Newsletter*, **35**, 27–31.

Ros, R. M. and Viñuales, E. 1992b, "Precisiones astronómicas con la fotografía," *Astronomia: Aspectos Científicos y Culturales*, Zaragoza University, 133–48.

Ros, R. M. and Viñuales, E. 1993, "Aristarco y las distancias al Sol y a la Luna," *Astronomía, Astrofotografía y Astronática*, Lèrida, **63**, 21.

Ros, R. M. and Viñuales, E. 1994a, "Graeco-Alexandrian astronomy today," *Manuskripte zur Chemiegeschichte*, Halle: TeaComNews, **5**, 49–58.

Ros, R. M. and Viñuales, E. 1994b, "El mundo a través de los astrónomos alejandrinos," *Astronomía, Astrofotografía y Astronáutica*, Barcelona, **69**, 12–15.

Ros, R. M. and Viñuales, E. 1996, "Relojes de Sol," *Astronomía, Astrofotografía y Astronáutica*, Barcelona, **82**, 34–9.

Ros, R. M. and Viñuales, E. 1997, *Planetas exteriores. Prácticas*, Madrid: Equipo Editorial Sirius, SA.

Ros, R. M. and Viñuales, E. 1999a, "Aristarchus' Proportions," in R. M. Ros, ed., *Proceedings of the 3rd EAAE SS*, Barcelona: Technical University of Catalonia, 55–64.

Ros, R. M. and Viñuales, E. 1999b, "Courves de Lumière de δ-Cephei et β-Persei," *Les Fiches Pèdagogiques du CLEA, hors sèrie no. 8*, Orsay, Francia.

Ros, R. M. and Viñuales, E. 1999c, "Evaluating the luminosity, temperature and percentage of the area covered of a solar eclipse," in R. M. Ros, ed., *Proceedings of the 3rd EAAE SS*, Barcelona: Technical University of Catalonia, 125–34.

Ros, R. M. and Viñuales, E. 2001, "Periodicidad del anillo de Saturno y las estrellas Cefeidas," *Enseñanza de la Astronomía*, Murcia, 31–33.

Ros, R. M. and Viñuales, E. 2001, "Puestas de Sol," *Enseñanza de la Astronomía*, Murcia, 35–7.

Ros, R. M., Viñuales, E., and Saurina, C. 1993, *La Fotografía: Una Herramienta para la Astronomía*, Zaragoza: Mira Editores.

Ros, R. M., Viñuales, E., and Saurina, C. 1995, *Astronomía: Fotografía y Telescopio*, Zaragoza: Mira Editores.

Viñuales E. and Ros, R. M. 1996, "Distances to the outer planets," in *Proceedings of the 5th International Conference Teaching Astronomy*, Barcelona, 191–3.

Viñuales E. and Ros, R. M. 1997, "Determination of Jupiter's mass," in R. M. Ros, ed., *Proceedings of 1st EAAE SS*, Barcelona: Technical University of Catalonia, 228–34.

Viñuales, E. and Ros, R. M. 1998, "An estimate of the number of stars visible to photographs," in R. M. Ros, ed., *Proceedings of the 2nd EAAE SS*, Barcelona: Technical University of Catalonia, 79–88.

Viñuales, E. and Ros, R. M., 2002, "Studying the ecliptic through photography," in *Proceedings of the 6th International Conference Teaching Astronomy*, Barcelona, 184–6.

Comments

Anonymous: We need to support the notion of using mathematical tools as concepts in linking astronomy to mathematics. Feteris and Hutton (1997) reported success in having students construct and use "trivial" astronomical tools in pre-service (tertiary) teacher training courses.

Reference

Feteris, S. and Hutton, D. 1997, "Astronomy laboratory: what will we make today?" *Publications of the Astronomical Society of Australia*, **17**(2), 116.

Julieta Fierro: Could you please give us the references of your teaching materials?

Rosa M. Ros: I published these papers in the European Association for Astronomy Education (EAAE) Summer Schools Proceedings, and they are available on the EAAE webpage (www.eaae-astro.org) and also in *Teaching of Astronomy in the Asian-Pacific Region*.

Open discussion

David McKinnon: Astronomy is an ideal integrative field. We alienate teachers enough by treating it as a "subject," one in which they feel that they have no "expertise." This is especially the case in primary schools where the teachers "teach" all of the "subjects" in the primary curriculum. Integration can happen by employing a thematic approach to the "teaching" of astronomy, which is driven by the students' interest.

Carlson R. Chambliss: I teach astronomy in a small university in Pennsylvania that has a planetarium. Pennsylvania is an unusual case due to the presence of Spitz Laboratories, the leading manufacturer of planetariums in the state. There are far more planetariums in Pennsylvanian secondary schools and colleges than anywhere else in the USA or elsewhere. High school planetarium directors usually do K–12 (6–17 year-olds) planetarium sessions.

Jayant Narlikar: By and large, astronomy in Indian schools is introduced as an appendage to geography. It hardly does justice to the scope of the subject or to the curiosity of the student. My experience with the numerous postcards I receive from secondary school students is that they have read a lot on the descriptive aspects of astronomy but would like to know the "why" behind them. As such I feel that O and A level physics will be a suitable stage when the "astrophysics" part could be introduced to the students.

Case Rijsdijk: We should be targeting teacher training institutes – knowledgeable teachers will use astronomy in the curriculum.

Maher Melek: I think that while one is building an integrated curriculum for astronomy with practical, theoretical physics, and mathematics, one should take into consideration clarification of the following issues:

- the observational facts (descriptive);
- the tools that are used in astronomy (either observational or theoretical);
- the existing theories that help us to understand and analyze different astronomical structures and phenomena.

Lorraine Mencinsky: The use of amateur astronomy groups in the school environment is a very valuable and under-used resource. Teachers are not aware of the depth of knowledge and range of equipment that amateurs can contribute. Local amateur societies are very willing to share their equipment for little or no reimbursement. Their enthusiasm is usually transferred to students and parents alike.

John Hearnshaw: Astronomy is a great medium for generating interest in physics and science in general. But for those who aspire to become professional astronomers, I emphasize again the need for a solid foundation in physics and mathematics. If we over-promote high-school astronomy, there is the danger of encouraging mathematically weak students to seek astronomy careers.

Bruce McAdam: Astronomy encompasses more than optical wavelength observation. I am a radio astronomer, and remind you that school observations can be made also with very simple and cheap radio equipment. The programs can exist in daylight, so the teacher doesn't need to organize night sessions. Students can observe meteor trails, Jupiter, the galactic plane, and the H I spectral line. Over the last five years, developments in Web communication also make it possible to observe in daylight during school hours using telescopes in time-zones where it is night time. We should start thinking globally – this should be easy for astronomers.

Vikram Ravi: Facilities (observatories) for astronomy in rural areas in Australia (i.e., the Australia Telescope Compact Array near Narrabri, New South Wales), are rarely used as part of astronomy classes because of the inertia of the teachers. Another reason might be that teachers seem always to be in a hurry to "complete the syllabus," resulting in few or no practical activities outside those specified by the syllabus, which are only designed (in my experience) to teach you something, not to interest you.

Pierre Encrenaz: For radio astronomy, there is a kit using a TV satellite antenna with a "Radio Shack" amplifier and correlator. The overall cost is around 10,000 euros. All the information can be obtained from Kathy Morellou at Onsala Space Observatory at Chalmers University.

Jay Pasachoff: A similar project exists at Haystack and the cost is US $3,000–$4,000. We should make sure that these people communicate with each other.

Connie Walker: It is necessary to have opportunities, knowledge-building and resources available for teachers, but also to have continued support for teachers, to retain the young teachers, and to renew the interest of the veteran teachers. The Teacher Leaders in Research-Based Science Education program at the National Optical Astronomy Observatory in Tucson, Arizona, is an example of this. Some statistics: out of 600 incoming pre-service teachers to be trained as science teachers only about three will be teaching after five years.

Reference

http://curie.umd.umich.edu/TeacherPrep/52.pdf

3

Engaging gifted science students through astronomy

Robert Hollow

Formerly at Tara Anglican School for Girls, now at
Australia Telescope National Facility, Epping, NSW, 2121, Australia

Abstract: Astronomy is a subject that poses many deep questions that intrigue students. It can effectively engage gifted and talented students in their school years. Numerous international and Australian schemes utilize astronomy as a means of challenging and extending such students. A variety of approaches include individual or mentored research projects, collaborative group tasks, distance-education courses, and classroom extension. Many schemes utilize access to online resources, communication tools, or remote telescopes. Several schemes are examined as case studies to highlight effective strategies. Some critical factors behind successful initiatives are identified, and implications for possible future schemes are discussed.

Astronomy is a subject that poses many deep questions that intrigue students. If presented in a relevant and stimulating manner it can effectively engage gifted and talented school students. Numerous international and Australian schemes utilize astronomy as a means of challenging and extending such students. The challenge is to learn from the successful schemes and build on them so that more students have access to them.

There is much debate in educational circles as to what constitutes a gifted student. However, Gagné's Differentiated Model of Giftedness and Talent is one that is widely used by educational bodies and so can serve as a means of definition. In this model (Gagné, 1996), gifted students have an *aptitude* in the top 15 per cent of their age peers in one or more of the following domains: intellectual, creative, socioeffective, sensorimotor and "others." *Talents* are skills (or abilities) and knowledge in one or more domains that have been carefully and systematically developed so that students perform in the top 15 per cent of their age group. Inherent in Gagné's model is the concept that being gifted does not necessarily lead to talent. Talent must be nurtured. Identification of gifted students is important if we are to help them move from the gifted to the talented. Gifted students may be identified through a variety of means: by teachers, parents or even themselves. Teacher observations are often the key method for gifted science and mathematics students (Stepanek, 1999). Diagnostic tests are commonly used at stages of schooling, while indicators such as participation in unusual hobbies or other interests can also be used. The aptitude domain most relevant to science is the intellectual one, linking to the academic and technology fields of talent, although this is not necessarily to the exclusion or detriment of other aptitudes or fields.

Given that a student is gifted in science, should we try to engage him or her in special programs or extension work? The answer to this is "yes," if such programs are structured to aid development from a gifted to a talented student. There are pragmatic reasons as to why this is desirable. Firstly, these students are potentially our future professional astronomers, scientists and engineers. Gifted students face considerable pressure on their fields of study and choice of careers. Unless presented with knowledge about and exposure to science and technology careers, students may opt for more traditional paths, often due to parental or economic pressures. Even if they do not aspire to a career in science, it is vital for society to have scientifically literate leaders, managers, and lawyers. Developing relevant and engaging learning experiences fosters positive views of "Science." It should also result in more positive interactions in the classroom and at school, with improved educational outcomes overall. Some gifted students exhibit behavioral problems in class, but this often arises because of their boredom with the simple, repetitive work presented to them. Another less measurable but equally worthwhile reason for engaging gifted science students in special programs is that, when students experience a breakthrough or "wow" moment, they can internally connect several strands of knowledge to explain some event or observation. This "thrill of discovery" (Verschuur, 1990) can have a significant and lasting impact on students.

Astronomy is a discipline that lends itself to engage gifted students and help develop their skills so that they emerge as talented students. It tackles "big questions" that many students find challenging or even confronting. Students of all ages may be engaged by it. Successful programs for gifted school students require an emphasis on learning concepts and higher-order thinking skills (Stepanek, 1999). Problem-based learning activities incorporating the use of technology as a learning tool can be used to model the scientific and investigative process. Context-rich learning examples abound in astronomy and can be used to develop students' skills in problem-solving, mathematics, science, literacy, and information and computing technology. An effective program will utilize students' existing knowledge, skills, and talents, yet also provide scope for developing new ones.

A wide range of programs and activities in astronomy are already developed that are suitable for students at all stages of schooling, K–12.[1] Relevant approaches include:

(1) one-off events such as a guest speaker;
(2) extension work within the normal class environment or by withdrawal within the school;
(3) extra-curricular opportunities such as a school astronomy club;
(4) gifted and talented camps, summer school programs, or their equivalents;
(5) school-district, system, state, or national curriculum schemes;
(6) online or distance education programs or subjects;
(7) early admission to a university subject;
(8) student research projects;
(9) combinations of the above.

One-off events such as a visit by a scientist can make a positive impression on a child and may serve as an initial source of inspiration, but unless they are followed up in some manner they can be a wasted opportunity. They are more likely to be of long-term benefit if incorporated into a larger program. This is not to undervalue the importance of such events, however, especially if they arise out of unplanned or unforeseen events.

In-school extension schemes can be a very effective method of engaging a streamed group or individuals but require considerable teacher enthusiasm, expertise, and a supportive, flexible school environment. An example would be a top-streamed class completing the syllabus requirements early and then working on a different task or more detailed work on the syllabus topic. In 1997 I had such a Year 10 class at Blue Mountains Grammar School, an independent,

---

[1] K–12: beginning of elementary school to the end of secondary school.

co-educational K–12 school 100 km inland from Sydney. We developed and ran a four-week topic on multi-wavelength astronomy that extended some syllabus points. The focus of this program involved a three-day visit to the 22-m Mopra radio telescope operated by the Australia Telescope National Facility. Here, the students were able to operate the telescope and obtain data under the guidance of professional astronomers from University of Western Sydney (UWS). The immersive experience of this program was valuable, with three of the group choosing to use the telescope later for their assessed student research projects that formed part of their matriculation mark in physics.

Some schools run extension programs, often for small groups of students, on a withdrawal basis from their normal subject lessons. Care is needed with this approach to avoid the perception that being gifted requires students to do more work than their peers. Ideally they may do different work but should not be required to do all the simpler or repetitive work that other students may do.

Specialist camps or summer schools can provide effective, targeted programs. A successful example that has run for over 40 years is the Summer Science Program, SSP (Furutani, 2001). This successful American scheme is project-based, with students working in teams to plan, conduct an investigation, and analyze data to determine an asteroid's orbital elements. It has a strong focus on mathematics, physics, and astronomy, plus an emphasis on social interactions for students. Now hosted at two sites in the USA, this six-week residential program also exposes students to practicing professional astronomers as lecturers and role models. Participants find themselves having to tackle real problems and find solutions under time constraints while learning to work cooperatively with others. Several alumni have gone on to hold important positions in professional astronomy.

The option of students taking one or more university courses before completing secondary education is available in some countries and states. While this may be valid for some students, others find the nature of many first-year courses off-putting because of large numbers in lecture theaters and the relative difficulty in engaging with lecturers in comparison to a school classroom. Perhaps a better approach pedagogically is to develop specific advanced courses that cater for gifted students within an education district or state curriculum. Such curriculum-based schemes exist in several countries. One example is the Cosmology Distinction Course, a matriculation-level subject that has run for the last ten years in New South Wales, Australia (Hollow, 1995). (See also the poster paper by Hollow *et al.* in this book on p. 120.) During this one-year course, students meet at two residentials, and complete assignments, exams, and a major project of their own choosing. The course is delivered by distance education, with provision for extensive student and staff communication via the Internet; it is open to gifted students who have accelerated in one or more of their matriculation-level subjects at least one year early. One particular value of this course is that it allows students from remote rural areas to meet with like-minded students from elsewhere in the state and keep in touch with peers and lecturers using the Web-based tools that were developed by one of the course students (McKinnon and Nolan, 1999). The residential component encourages students to interact socially with each other, meet practicing researchers, and tour a range of observatories and telescopes. A more recent online education initiative is the Cyber Astronomy Course run by Hong Kong University of Science and Technology for senior-high-school students (Chan and Wong, 2003). Together, such schemes show the power of computer technology in fostering student communication and course delivery. As a 2002 student of the Cosmology Distinction Course commented:

It gave me a glimpse of an entirely different (real?) way of learning which was far more involving than classes at school, i.e. having the onus put on the student to . . . direct their own learning rather than a set of criteria and dot points to memorize.

One key method of engaging students and fostering skill development is that of the student research project. In getting students to tackle a research project we are asking them to actually do some science rather than just repeat textbook "experiments." Evaluation of the impact of student research schemes shows strong links to more positive student attitudes towards science and technology, more students going on to engineering and science courses, and strong commitment from teachers, students, and employers as to their worth (Woolnough, 2000). Such projects naturally require students to apply analysis, synthesis, and evaluation skills, the highest-order skills under Bloom's Taxonomy of Educational Objectives. Several schemes now running or under development provide students with access to sophisticated telescopes with charge-coupled devices (CCDs) allowing them to obtain real and worthwhile data (Beare *et al.*, 2003; Smith *et al.*, 2001). These include the following facilities for optical observations.

(1)   Hands-On Universe (HOU);
(2)   Telescopes in Education (TIE);
(3)   The Charles Sturt Remote Telescope;
(4)   The Faulkes Telescope Project;
(5)   National Schools Observatory;
(6)   Bradford Robotic Telescopes.

There are also equivalent schemes for radio astronomy such as the Goldstone Apple Valley Radio Telescope Project, GAVRT (www.gavrt.org). The data obtained can be used for projects ranging from simple one-off observations to more advanced, long-term ones. Access to such instruments can provide a strong and effective motivator for students and teachers. Some schools have even developed powerful telescopes and research programs themselves, often with some outside professional support. Perhaps the best example of this is the Taunton School (Hill, 1995) in the UK, which has built a range of radio telescopes and observed a range of sources. There is considerable scope for institutional support and professional mentors for groups tackling projects (Kadooka, Meech, and Bedient, 2002), and this can also foster beneficial links within a local educational community. Some projects also lend themselves to local or global collaborations among student groups. This adds to the educational value of the task.

A common mistake for teachers involved in setting up such student research schemes is to artificially impose a limit on students' abilities and efforts. Gifted and talented students have a habit of rapidly exceeding our preconceptions. This does not mean, however, that there is no place for a structured approach. The use of a scaffold of concepts and task requirements for students provides guidance and support and increases the educational value of the project. The Internet provides an effective means for dissemination of exemplars of student projects; indeed journals such as the *ScI-Journal* (http://www.sci-journal.org/) already exist for this purpose. Individual astronomy research projects for students, such as HOU, also provide examples of student work (http://www.handsonuniverse.org/).

A scheme run for several years at Blue Mountains Grammar School (BMGS) in New South Wales saw numerous students tackle astronomy-based projects as part of their assessment for matriculation-level physics (Hollow, 2000). Topics included:

(1) naked-eye meteor observations;
(2) differential CCD photometry of short-period variable stars;
(3) differential CCD photometry of quasars;
(4) comparison of cluster ages using color-magnitude diagrams;
(5) mm-wave observations of Orion A;
(6) Investigations of Large Magellanic Cloud (LMC) regions N159 and N160 using data from the Australia Telescope Compact Array.

The student who did the last project won the national prize for student research for her project.

Follow-up surveys of some of the students involved indicated several positive outcomes. A high percentage of these students continued in science/technology study and career paths. Even those who went into other fields commented favorably on their research project as a mechanism for improving their information and computer technology skills, and their time- and project-management skills. All gained in confidence and had a positive attitude towards science. Some had benefit in terms of scholarships for university entry.

Negative aspects included conflicts with the demands of other subjects, the fact that research skills were not "examined" in the external final examination, and the heavy time commitment from both students and teachers. Resourcing is also a concern, and the success of the BMGS scheme was due in no small part to the support and facilities provided by the University of Western Sydney. Extending schemes such as this to other schools could exert excessive pressure on already overstretched tertiary institutions.

Overall, those programs that are of most benefit for gifted students are those that:

(1) emphasize learning concepts;
(2) develop higher-level thinking;
(3) promote inquiry, especially problem-based learning;
(4) utilize technology as a learning tool;
(5) model the scientific process, using (experimental?) design procedures;
(6) encourage social interactions and development;
(7) lead to improvements in mathematical, analytical, and literacy skills;
(8) develop independent learning and time-management skills (Stepanek, 1999).

Student research projects, particularly those involving group work, can meet all of these conditions and are a powerful method of challenging and extending the gifted student.

If we accept the value of engaging gifted students, we need to consider what is necessary for us to move forward. First is the realization that we cannot afford to neglect or ignore gifted students. Educational evaluation of effective (and ineffective) programs is needed so that we can identify what are the key factors. These then need to be developed, modeled and their use extended. Doing so requires strong linkages between professional astronomy and educational groups so that engaging and relevant programs can be developed. We do not want to underestimate student ability or potential; therefore we need to develop scaffolded programs with potential for expansion and ongoing development. Teacher training and ongoing support to raise awareness, confidence, and skills will be critical and requires a strong commitment to resource allocation and time.

Astronomy provides an effective means of engaging and challenging our gifted students. Relevant programs can help their development into talented students who possess a positive attitude towards science and a deeper understanding of its processes and value.

References

Beare, R., Bowdley, D., Newsam, A., and Roche, P. 2003, "Remote access astronomy," *Physics Education*, **38**(3), 232–6.

Chan, C. W. and Wong, K. Y. 2003, "Cyber Astronomy: a cyber university course for school students," *Physics Education*, **38**(3), 237–42.

Furutani, T. 2001, "Asteroids, teenagers and real science," *Sky and Telescope*, **101**(3), 76–9.

Gagné, F. 1996, "A Thoughtful Look at Talent Development," accessed at http://www.coe.unt.edu/gifted/Resources/gagne.htm, on 14/10/03 on Internet Explorer 6.

Hill, T. 1995, "High frequency radio observations of comet Shoemaker-Levy 9," *Physics Education*, **30**(3), 135–42.

Hollow, R. 1995, "The Cosmology Distinction Course for gifted students," *Physics Education*, **30**(3), 129–34.

Hollow, R. 2000, "The student as scientist: secondary student research projects in astronomy," *Publications of the Astronomical Society of Australia* (ASA), **17**(2), 162–7.

Kadooka, M. A., Meech, K. J., and Bedient, J. 2002, "TOPS telescope projects on variable stars and other objects," *Journal of the American Association of Variable Star Observers*, **31**(1), 39–47.

Kiefer, W. S., Herrick, R. R., Treiman, A. H., and Thompson, P. B. 2003, "Exploring the solar system: a science enrichment course for gifted elementary school students," in *Proceedings of the NASA Office of Space Science Education and Public Outreach Conference*. http://www.lpi.usra.edu/science/kiefer/Education/expsolsystem.pdf

McKinnon, D. H. and Nolan, C. J. P. 1999, "Distance education for the gifted and talented: an interactive design model," *The Roeper Review*, **21**(4), 320–5.

Smith, D., Penston, M., Roche, P., and Murdin, P. 2001, "Beyond backyard astronomy: professional telescopes online for schools," *Physics Education*, **36**(3), 178–83.

Stepanek, J. 1999, "Meeting the needs of gifted students: differentiating mathematics and science instruction," Northwest Regional Educational Laboratory http://www.nwrel.org/msec/images/resources/justgood/12.99.pdf.

Verschuur, G. L. 1990, "The thrill of discovery," in J. Pasachoff and J. Percy, eds., *The Teaching of Astronomy*, Cambridge: Cambridge University Press, 68–71.

Woolnough, B. E. 2000, "Authentic science in schools? An evidence-based rationale," *Physics Education*, **35**(4), 293–300.

Comments

Maher Melek: Have you noticed some correlation between gifted students in the sciences and their interests in other stuff like art, music, etc.?

Rob Hollow: There are several studies that show that many (but not necessarily all) who are gifted in science are also gifted in other areas like music. From my own involvement with gifted students in the Cosmology course over 10 years, we have had many students who have displayed exceptional linguistic or musical skills. I am sure that some of these may have also been strong at visual arts, but I was not able to assess this. At some of the residentials, however, we have had recitals by the students that were of outstanding quality.

Reference

http://www.nrc-cnrc.gc.ca/newsroom/speeches/convergence_e.html

Bill Zealey: It's great to motivate students at high school, but we have to pick up the thread at university. Students who have done projects at school can get turned off by standard first year physics at university. At Wollongong we ensure students have an option to do project work in their first year laboratories.

Rob Hollow: Yes, we've encouraged students to continue with project-based work over vacation and at university with Maria Hunt. Several have gone on to complete degrees in astronomy or astrophysics.

Jay Pasachoff: See also:
http://www.teachspacescience.org
http://www.solarsystem.nasa.gov/educ.

Poster highlights

The topic of including astronomy as part of a general curriculum is first addressed by **Nassim Seghouani** in a poster entitled **The project of teaching astronomy in Algeria**. Seghouani mentions that there is currently no teaching of astronomy in Algeria despite a glorious history in this domain. The Algiers Observatory, constructed in 1890, participated since its creation in many international projects, and it was the leading observatory in the international project "Carte du Ciel." Today, and since 1985, research in astronomy in Algeria has started up again, and the country has firmly decided to invest in this domain. New instruments have been acquired lately, such as an 80-cm telescope. The introduction of astronomy and postgraduate astronomy courses is currently in progress, along with wider aspects attached to this project. The organization of summer schools as well as the introduction of astronomy teaching in secondary schools is also discussed.

Further global perspectives are provided by representatives from many of the countries created upon the dissolution of the former USSR, including a paper by **Elchin S. Babayev**. Babayev and his colleagues describe the current situation of astronomy education in Azerbaijan. Azerbaijan has a long history of contributing to astronomy, and has an under-utilized astronomical treasure in the Maragha Observatory, located in the south of Azerbaijan, originally established in the Middle Ages by astronomer Nasiruddin Tusi.

They examine the process of teaching astronomy and space science education in Azerbaijan as carried out in high schools, gymnasiums, universities, and other academic institutions through lectures, seminars, and conferences. Obligatory teaching of astronomy in high schools, gymnasiums, lyceums (three kinds of secondary school), as well as the teaching of astronomy and space sciences in many university departments, gives special status to astronomical education in the State Education Curricula and allows people to become familiarized with astronomy by involving them in deep astronomical study with contemporary research.

They examine effective learning and teaching techniques, the popularization of space and astronomical knowledge, and an overview of astronomy in the schools in Azerbaijan. In addition to classic education activities in schools and universities, the specialized community in Azerbaijan invests a lot of effort in general public outreach.[1]

Muhamed Muminovic recounts the production and evolution of Bosnia-Herzegovina's first general astronomy textbook in **The New Bosnian Textbook of Astronomy**.

---

[1] Further perspectives on astronomy education in developing countries can be found in Part VII.

In 2002–3, Muminovic's textbook titled *Astronomy* was prepared. In this book, he tried to keep the spirit of his previous editions, starting in 1972, with the first printed text on general astronomy in Bosnia-Herzegovina. After five editions, the sixth one is a totally new book. In comparison with the fifth edition, the text of the new *Astronomy* has not drastically changed. But, using computer graphics, especially computer air-brushing techniques (by Bosnian artist Mr. O. Pavlovic), they produced something new along the lines of similar astronomical textbooks elsewhere in the world. Explanations of some astronomical phenomena that can't be described very easily with words, or are such that we don't have direct images of them, are much easier to explain with color images. The new textbook and techniques can be of interest to others who are involved in astronomical teaching and popularizations. *Astronomy* represents a huge step forward for astronomy in Bosnia-Herzegovina.

Perspective from the Caribbean is provided by **Derick Cornwall** in a poster entitled **Astronomy education – the Caribbean experience**

This paper presents the results of a survey conducted in high schools among teachers and students to establish the status of astronomy education in the Caribbean. 524 students were sampled representing 10 schools across urban and rural areas of the country; 66 per cent of the participants were female and 34 per cent male. The majority of the students, 42 per cent, fell in the age group of 14–16 (the upper level in school). Of all the respondents, 79 per cent indicated that they were interested in astronomy, and 65 per cent said that they would like to study it in school. (Presently it is not formally taught, except as an option in some schools.)

The survey indicates strong interest in astronomy among students and educators alike, but because of inadequate resources, and a lack of teacher training, it has not been given the attention it deserves. 98 per cent of students questioned indicated that there was no astronomy club at their school, and 73 per cent said they would have liked to have an astronomy club at their school. Of the respondents, 77 per cent had never looked through a telescope. On testing their general knowledge of astronomy (by multiple choice), it was found that 87 per cent could not distinguish between astronomy and astrology, 76 per cent did not know what a constellation was, 74 per cent could not identify the largest planet in the solar system, and 68 per cent did not know we lived in a spiral galaxy. 46 per cent of the students believed in UFOs, while the rest did not.

In addition, Cornwall presents the results of the Caribbean Institute of Astronomy's (CARINA) ongoing efforts at astronomy education and popularization, both independently and in conjunction with the only science center in the Caribbean. Problems specific to astronomy education in the Caribbean are also discussed with regard to funding, communication, and a lack of human resources.

A representative from the Island of Mauritius, **Nalini Heeralall-Issur**, writes about **The teaching of astronomy at the University of Mauritius**.

Heeralall-Issur describes the present teaching of astronomy in Mauritius. Astronomy is taught mostly at the university level as electives (Astrophysics I and II) within the B.Sc. Physics course or the M.Sc. Physics with Astrophysics Specialization Option. However, there are limitations due to both facilities available and staff resources at the higher level. At the lower level, very few secondary school students are taught astronomy, and even many well-educated adults have no basic knowledge of astronomy.

Some realistic suggestions for improving the teaching of astronomy at all levels in Mauritius are discussed.

From India, **V. B. Bhatia** discusses **Astronomy in Indian schools**.

The tradition of astronomy in India goes back to ancient times. Indeed, many festivals and rituals are associated with astronomical phenomena. Indian children start learning the rudiments of astronomy early on in primary school, but primary teachers are often not equipped to handle the subject. As a result, students often come away with incorrect information. Their first serious contact with astronomy occurs in class X when they are 15 years old. Until 2002, astronomy was represented in class XII as well, but it has now been dropped. This is a serious setback for the study of astronomy in India.

In class X, astronomy forms part of a general science curriculum. Since children at this stage are not proficient in physics or mathematics, the subject remains descriptive, though there are useful activities for the children to do. However, the teachers are not equipped to handle this subject, and there is no help in the form of visual material. The subject thus remains neglected.

The Indian astronomical community can help by training teachers and providing visual material. It must also urge authorities to reintroduce astronomy in class XII if astronomy is to flourish in India. Moreover, India needs to network with developing countries, share experiences with them, and evolve a strategy that promotes astronomy.

Discussing all levels of astronomy education in Russia, **Alexander Tron** presents a poster entitled **Astro-education in St. Petersburg: from school to postgraduate**.

Astronomy teaching in St. Petersburg has a long history. During the last 30 years (and currently), it is presented mainly by structures additional to the high-school curriculum (astronomy as a separate subject is taught in about 7 per cent of the St. Petersburg schools), such as: Youth Astronomical-Geographical School "Earth and Universe," Astronomical Olympiad, and classes of the St. Petersburg University Gymnasium. All teaching is provided by professional (mainly young) astronomers, not by schoolteachers, and mainly on a volunteer basis. There are about 100 high-school students involved annually. For four years (from age 13 to age 17 – graduation from high school) students take courses in basic cosmography, astrometry and celestial mechanics, astrophysics and cosmology, geophysics, advanced topics in mathematics and physics, history of astronomy, and space research.

Computers and the Internet are intensively used, though there is still a great shortage of hardware and access to the Internet, mainly for financial reasons. The observational base is poor too, for the same reasons, though the Pulkovo Observatory and the Astronomical Institute of the St. Petersburg University are trying to help. There is also cooperation with the "Hands-On Universe" program. Half of the annual enrollment of the students of astronomy in the St. Petersburg University come from these systems of education.

A representative from the Near East, **Khursand I. Ibadinov**, explains the state of **High and higher astronomical education in Tajikistan**.

In all three types of high school in Tajikistan, astronomy will be studied from 2003 onwards; however, there are no educational astronomical observatories. At many schools there are no astronomical instruments or modern manuals. In 1999, the Tajik State National University (TSNU) instituted astronomy as a specialty of study, and the Department of Astronomy was organized. In the four courses of this specialty, 95 students are now enrolled. In 2003, the first graduate in the bachelor program graduated.

Ibadinov also recounts some of the other difficulties faced at the TNSU: there are no astronomy textbooks written in the Tajik language, and there is no observatory at TNSU.

In 2002 at the Institute of Astrophysics of the Tajik Academy of Sciences, a Specialized Council for awarding a scientific degree in astronomy was established. Thus, in Tajikistan, the conditions for obtaining a solid astronomical foundation and obtaining a scientific degree in astronomy are slowly being built. The astronomical education in Tajikistan needs modernization.

Salakhutdin Nuritdinov discusses similar problems in **Astronomy teaching in the schools of Middle Asia**.

Nuritdinov gives an overview of the state of astronomy in today's schools (elementary and secondary) in Middle Asia, specifically the countries of the former Soviet Union. He also looks at the following issues in detail:

- the common state of teaching of astronomy in Middle Asian schools;
- the school curricula, comparisons of some subjects, and the necessity of teacher training in the speciality "Physics and Astronomy";
- the evaluation of what constitutes effectiveness in teaching astronomy;
- the textbooks on astronomy that were written during the days of the former Soviet Union, which contain cumbersome and complex fundamental processes and are not effective;
- maximizing effective learning of astronomy by including in textbooks the names and scientific results of well-known Asian astronomers from the Middle Ages (Al-Fragan, Al-Biruny, Al-Chagmany, Ulugh-Beg, Ali-Qushchi, etc.)

Nuritdinov continues with another poster on **Possible collaboration perspectives in astronomy education** and reminds us that IAU Commission 46 often concerns international educational collaboration. As more astronomers graduate from Uzebekistan universities, it becomes necessary for them to discuss collaborative perspectives.

In countries belonging to the former Soviet Union, educational institutions are starting to train students for bachelor's and master's degrees. It is imperative that we begin working on consistent educational curricula. Bachelor programs must contain 60–70 per cent content that is identical to that of other countries so that students can continue with graduate school in other countries. The remaining 30–40 per cent can consist of interests particular to that country.

We should look towards creating an international virtual library of astronomy resources so students can have access to textbooks and scientific journals necessary for study. We should consider organizing summer and winter session schools each year for traveling students. In Uzbekistan, we did not receive timely word of this conference because the only school able to train astronomers is the University of Central Asia.

Lastly, **Alisher S. Hojaev**, also from Uzbekistan, presents a poster on **How to increase correct learning of astronomy**.

Astronomy is the most popular and attractive of the sciences. Undoubtedly, the importance of exact knowledge of astronomy is becoming more and more crucial.

However, the level of teaching of astronomy at schools, lyceums, colleges and even universities, especially in the countries of the former Soviet Union, is far from satisfactory. Moreover, some astronomy-related information in the mass media (TV, newspaper, etc.) often contains incorrect information.

Astronomical organizations and individual astronomers are tremendously important in the application of correct knowledge to the curriculum, and in the popularization and control of the astronomical presentations in the media.

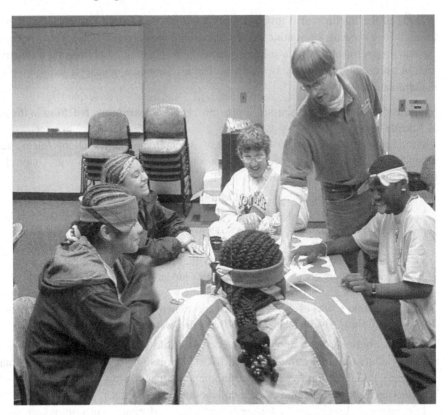

Happy high-school students engaged in hands-on astronomy activities. Photo by Mary Kay Hemenway.

Increasing access to the Internet, a fairly recent event in Uzbekistan, gives new possibilities for improving astronomy education, and utilizing remote education tools to open new frontiers for the popularization of astronomy. The arrangement of practical excursions, public lectures and presentations of new concepts in astronomy, as held at Ulugh Beg Astronomical Institute, is discussed. Some other ideas to improve astronomy education are given.

Part II

Astronomy education research

Introduction

There is an interesting cartoon that shows two small boys and a dog. In the first panel, one boy says to the other "I taught Stripe (the dog) how to whistle!" In the second panel, the other boy replies "I don't hear him whistling!" In the third and final panel, the first boy says "I said I taught him; I didn't say he learned it."

This cartoon emphasizes the difference between teaching and learning. Research shows that, in the most traditional methods of teaching, the amount of learning may be zero. Of course, there is more to teaching than the learning of facts. In fact, there is an old saying that "education is that which remains after the facts have been forgotten." So it can include the intangible effects of an inspirational teacher.

There are many important challenges in the effective teaching and learning of astronomy, and most of them are amenable to research. One challenge is students' deeply rooted misconceptions about astronomical topics; non-expert teachers often share the same misconceptions. Some of these misconceptions are caused by the influence of religion or of popular culture. Others result from the three-dimensional nature of many astronomical concepts, or from the problems of moving from an observer-centered frame of reference to an external one, or from the enormous astronomical scales of size, distance, and time. Many of these concepts are intrinsically difficult, and the work of Piaget and others seemed to show that these concepts require students to have reached the appropriate stage of intellectual development – secondary school, for instance. Piaget's work has been widely challenged, and many of its supposed applications use oversimplified versions, and overgeneralize its applicability. Here, too, more widespread understanding of the status of contemporary education research would be helpful. Further, different people learn in different ways. The question of teaching or engaging gifted minorities separately from average or below-average students is a raging debate. Must everything we teach be for the lowest common denominator?

Some concepts are intrinsically boring, and day and night, seasons, moon phases, eclipses, and tides can fall into this category. Though each of us isn't young Isaac Newton, even the fall of an apple can be inspiring if it reaches a mind prepared to consider its consequences. Finally, there is the problem that much hands-on (or eyes-on) astronomy must be done at night, and "the stars come out at night; the students don't." The breakthrough of a British organization building and operating a telescope in Hawaii, so that British students can control a night-time telescope during their daytime, may be a harbinger of the breaking of old boundaries with the help of the Internet. Further, more advantage can be taken of the daytime for observing our interestingly changing sun, in part with the recent availability of inexpensive and easy-to-set-up solar telescopes.

The apple at Isaac Newton's (1642–1727) feet is red in this fiberglass statue at the Inter-University Centre for Astronomy and Astrophysics in Pune, India. Photo by Jay Pasachoff.

Astronomy teachers around the world have dealt with these issues. Many of them have shared their successes (or failures) with colleagues, in papers in journals and magazines, and in the proceedings of conferences and workshops. Education researchers have carried out formal studies of some of these issues, and published the results in refereed journals. Significant papers are therefore scattered through literally hundreds of publications. In Part II,

John Broadfoot and Ian Ginns, Janelle Bailey and Tim Slater, and Leonarda Fucili have provided us with a wealth of information and references. Other useful bibliographies are available on the websites of the American Astronomical Society (www.aas.org/education) and the Astronomical Society of the Pacific (www.astrosociety.org). Nevertheless, there is a need for a comprehensive searchable database. We hope that our own commission's website at www.astronomyeducation.org becomes a useful link among astronomy education efforts in all the countries of the world.

Sidney Wolff and Andrew Fraknoi describe an important step forward: the *Astronomy Education Review* – a free electronic journal that includes refereed papers as well as news and notes about astronomy education.

But there is much to be done. Although there is a great deal of research on teaching and learning about solar system concepts, including positional topics such as moon phases and seasons, there is little research on teaching and learning about stars, galaxies, and the universe. That may be partly because few courses progress beyond the solar system. Some of this restriction may come because of limited understanding of astronomy, and even limited knowledge of advances in the post-Piagetian psychology of education. Some comes from a lack of preparedness in science on the part of schoolteachers. Some comes from well-meaning committees in which astronomy and astronomers have not been well represented. Some comes from an attempt to limit teaching to only subjects that can be understood or experienced by all children. We hope that books and conferences like this one help broaden astronomy education in knowledgeable and appropriate ways.

4

Astronomy education research down under

John M. Broadfoot and Ian S. Ginns
Centre for Mathematics, Science and Technology Education, Queensland University of Technology, Australia

Abstract: A review of research conducted in Australasia into students' concept development in astronomy identifies a number of learning difficulties pertinent to astronomy, such as the lack of prior knowledge, intuitive or naive beliefs, juxtaposition and frames of reference, difficulties in mental-modeling, the inappropriate use of analogies, and the absence of consideration of the historical development of astronomy in teaching. The impact on, and effectiveness of, some teaching and learning strategies, developed and employed to address some of these learning difficulties, have been examined. A number of recommendations are proposed for teaching strategies and further focused research into the learning and conceptual needs of students.

4.1   Introduction

The curriculum and the teaching of astronomy in any school are driven by the syllabus and the individual school's developed and implemented curriculum. Other factors that may affect the delivery of the intended curriculum will include the competence and interests of individual teachers and the use of textbooks and other resources, to support this curriculum. The successful teaching of astronomy in primary and secondary schools is dependent on the teachers' own understanding of concepts and abilities to challenge students' prior conceptions, therefore the courses undertaken by pre-service teachers must be in the context of teaching astronomy, and teachers' own studies must challenge and develop their own understanding of concepts in astronomy.

To understand fully the possible derivation of alternative conceptions or difficulties in understanding key concepts in astronomy, it is essential to trace the stages of teaching and learning in astronomy through the curriculum. It may be that students have not been provided with the opportunities to challenge and/or restructure their prior knowledge or intuitive beliefs through appropriate teaching strategies. Teaching approaches need to incorporate strategies that engage students in challenging their prior beliefs and intuitive ideas, practicing juxtaposition and frames of reference problems, and constructing acceptable models of celestial phenomena. Such strategies might incorporate or re-enact historical discoveries (Noble, 1999), thus engaging students in thinking about astronomical phenomena from an intuitive position. These factors are discussed in terms of their implications for student learning, examples of strategies and associated difficulties faced by students, and possible recommendations for teaching and further research.

4.2 Curriculum in Australia

4.2.1 Tertiary courses

In the main, astronomy subjects belong to, or are serviced by, the physics, astrophysics, or mathematics departments (or schools) of tertiary institutions in Australia. Out of 30 tertiary institutions surveyed within Australia (Astronomical Society of Australia, 2000), only three offered an astronomy subject from outside the physics, astrophysics, or mathematics departments. Ten institutions offered their first-level astronomy subject to non-science majors, and only one institution offered a full semester subject designed specifically to train teachers to teach astronomy. The focus of many of these courses is on modern astronomy that includes concepts involving a high level of understanding in mathematics and physics. Much of the content of these courses does not appear to provide the pedagogical approach needed for the training of primary and secondary schoolteachers. Further research is needed to ascertain the extent of the inclusion of teaching and learning strategies to enable students to understand the fundamental interactions within our universe as seen from an egocentric viewpoint.

4.2.2 Primary and secondary syllabuses

All Australian science syllabuses are designed for the continuum of science education from Years 1 (or K) to 10, the compulsory years of schooling. A review of these science syllabuses of Australian States and Territories indicates that there is limited time devoted to topics in astronomy. In some of the syllabuses the topics are quite prescriptive and comprehensive, while in others, the lack of prescription may lead to limited guidance in the teaching of astronomy topics. Within all syllabuses there is opportunity to scaffold student learning through the totality of the syllabus. However, in most instances examined by the authors, this learning may not be continuous, resulting from a lack of sequencing in prescribed astronomy topics and limited opportunity for students to observe and explain from one year to the next. Constraints of time within schools, the pressures of covering other topics, limited resources, and the personal expertise or interests held by teachers, may lead to avoidance of the teaching of astronomy topics. More research is needed into the effect of teacher competencies in, and attitudes towards, the teaching and learning of astronomy at the primary and secondary levels.

4.2.3 Influence of textbooks on the curriculum

There are a large number of science textbooks on which most teachers are dependent to deliver the intended curriculum. In most of these textbooks the quantity and breadth of astronomy topics does not match the syllabuses. Support resources are often needed for the teaching of astronomy to comply with the syllabuses. The lack of sequential development of concepts in three major science textbooks is notable. For example, in one series (Stannard and Williamson, 1999), the sequencing progresses from Year 7, where students study the sun–Earth–moon system and the motion of the Earth, to observations of the night sky in Year 8. Another recent textbook series (Thickett, Stamell, and Thickett, 2000) begins with an egocentric viewpoint, with students in Year 7 looking for patterns in the sky and the planets (wanderers), but then focuses on contemporary astronomy and the exploration of space in Years 9 and 10.

4.3 Research into learning in astronomy

There are many unresolved teaching and learning difficulties that may be somewhat peculiar to astronomy (Taylor and Barker, 2000) including difficulties in visualization, mental modeling, and conceptual restructuring; however, there has been limited focused research in the specific area of astronomy teaching (Treagust and Smith, 1989; Taylor and Barker, 2000). Isaac Watts (Brooks, 1987) was one of the earliest psychologists to identify that the three-dimensional nature of astronomy challenged one's mind.

4.3.1   Visuospatial abilities

Ekstrom *et al.* (1976) defined two factors comprising spatial ability – spatial orientation and spatial visualization – as follows: spatial orientation – the ability to perceive spatial patterns or to maintain orientation with respect to objects in space (p. 149); and spatial visualization – the ability to manipulate or transform the image of spatial patterns into other arrangements (p. 173). Research studies in astronomy indicate that students require high levels of spatial ability thinking (Glynn, Yeany, and Britton, 1991; Hill, 1990; Vosniadou, 1991a, 1991b; Vosniadou and Brewer, 1987, 1989), and some measures of student spatial abilities, namely spatial orientation and spatial visualization, are predictors of students' abilities to succeed in solving problems focusing on the dynamic and three-dimensional nature of the universe (Broadfoot, 1995). Research into these aspects of spatial thinking, as well as mental modeling, are examined in the following sections.

Spatial orientation

Astronomy frequently requires students to imagine objects from other viewpoints (Broadfoot, 1995; Rock, Wheeler, and Tudor, 1989), which demands a shift in perspective or change in the student's frame of reference (Hintzman, O'Dell, and Arndt, 1981; Finegold and Pundak, 1990). A number of cognitive psychologists (Yackel and Wheatley, 1990; Kosslyn *et al.*, 1988) have also reported the importance of image formation in the learning process. Students need an absolute reference point before generating images (Howard, 1982); therefore, prior imagery experiences are important in students' concept development in astronomy, and unique individual experiences and logic are critical barriers to understanding in astronomy learning (Sadler and Luzader, 1990). In a study by Broadfoot (1995), students were observed to have difficulties with orientation in space, and location and identification of objects in space in relation to other celestial objects. For example, imaging of constellations in space with large areas of black background may pose problems for students in forming relationships between stars (Broadfoot, 1995). The relative location of objects that appear close in space can be correctly processed provided no perspective shifts are demanded. Orientation-bound frameworks, up and down, left and right, front and back, and orthogonal directions (Humphreys, 1983), are essential for understanding the relative positions and motions of celestial objects, which provide a basis for absolute and relative orientation. Individuals may initially comprehend the relative positions of objects such as constellations in space without a comprehensive orientational or orthogonal framework but fine-tune the exact location of objects in space with increasing field experience (Feteris and Hutton, 2000; Broadfoot, 1995). Interviews with, and observations of, students during astronomy field work supported the need for spatial training that involved shifts in student viewpoints of the type encountered in the recognition and orientation of groups of stars or constellations (Broadfoot, 1995). This developmental process of linking groups of stars to locate other celestial bodies may also contribute to the development of orientational frameworks.

Spatial transformations

Problem-solving exercises, dealing with orbital and rotational movements of celestial objects, frequently require students to interpret information from given data and then restructure the information into different viewpoints both verbally and diagrammatically (Broadfoot, 1995). Shepard and Cooper (1982) reported that orientation of images in three dimensions increases in difficulty as the angle of rotation of an object from the original position increases, and it becomes increasingly difficult or time-consuming for subjects to recognize the similarities between two-dimensional drawings of the same three-dimensional objects. However, in some cases, the recognition of objects from different viewpoints may not involve rotation (Pinker and Finke, 1980). Some young children may be incapable of creating a view from another position relative to their own (Piaget and Inhelder, 1956).

However, Takano (1989) has reported that the hypothetical viewer in his/her egocentric framework may be able to construct an internal representation of a form with or without the need for mental rotation through the linking of conceptual schema to orientation-based descriptions. In contrast, Marr (1982) proposed that objects can be described with reference to a set of principal and subsidiary axes without reference to any external framework and, further, that no mental rotation is needed to make comparisons between objects. Just and Carpenter (1985) also found that mental rotation was not needed for orientation-free descriptions, whereas mental rotation is needed when objects are orientation-bound. Whether students do locate and observe celestial objects using reference axes may have some bearing and may describe the mental processes used by some students when conducting field activities in astronomy.

In studying the sun–Earth–moon system, directional coordinates are necessarily involved. However, an object-centered viewpoint necessitates considerable mental rotation by the observer from his/her viewpoint. This has implications for the development of astronomical perceptions creating a possible conflict of object-centered (heliocentric) views with observer-centered (egocentric) views. Quite often the image viewed through a telescope is inverted and/or flipped left to right. Students have difficulties comparing and recognizing features or landmarks on the moon's surface because of extremes of orientation involving rotations in two directions: left to right and top to bottom. It is possible that students use landmarks, such as craters or maria, as a strategy to recognize the orientation of the moon's virtual telescopic image when compared to the naked-eye view (Broadfoot, 1995). Lucas and Cohen (1999) described the difficulties faced by students when relating heliocentric textbook diagrams about seasons to their own egocentric experiences and views. Sadler and Luzader (1990) also reported that students have difficulty in mastering spatial relationships posed in many textbooks. These findings have implications in situations where students need to transform textbook representations of the night sky or phases of the moon to three-dimensional views.

4.3.2 *Conceptual restructuring in astronomy*

Students come to class with "phenomenal" experience about the nature of celestial objects. Observational astronomy involves conceptual restructuring of prior knowledge and experiences about the shape, size, movement and location of celestial objects. Vosniadou (1989, 1991a) and Gentner and Stevens (1983) have identified three kinds of mental models – intuitive, scientific and synthetic – that may be used to explain observed phenomena and concepts in astronomy. The intuitive model is based on observational experiences of the natural world and requires little modification for accommodation of these observations. Scientific models support and agree with current scientific views. Synthetic models show a combination of intuitive and scientific views and represent some kind of misrepresentation of scientific information. Phenomenal experience of the Earth describes it as flat, stationary, much bigger than the sun or moon, and as the center of the universe. Most students accept the scientific model of a spherical Earth but this is in direct conflict with their own observational perceptions of a flat Earth. Observations of stars, the moon, planets and the sun apparently "revolving" around the Earth would also form an intuitive model, which is in keeping with the Ptolemaic Earth-centered model of the universe. This perception of the motion of the sun, moon, stars, and other planets rising in the east and setting in the west leads to an intuitive model that is based on the observer-centered percept.

From their intuitive models students construct mental models to solve problems in astronomy (Glynn, Yeany, and Britton, 1991; Vosniadou, 1989, 1991a; Vosniadou and Brewer, 1989) that provide insight into the constructs of students' knowledge bases. Vosniadou (1991a) concluded that a mental model of a spherical Earth is a prerequisite to understanding the scientific explanation of the day/night cycle, as is also an understanding of axis. These two concepts – spherical Earth and axis – are examples of interdependence among concepts that are very important in astronomy learnings. A flat Earth, stationary and central to the universe, effects of gravity, the sun and moon moving up and down or east to west, and the stars being small are just some of the intuitive models possessed by many students.

Many textbooks do not provide guided activities to facilitate the assimilation of increasingly abstract ideas represented by object-centered diagrams and descriptions for celestial objects and phenomena. Students need to recognize similarities and make links between the object-centered and egocentric viewpoints to facilitate assimilation and accommodation (Broadfoot, 1995). Vosniadou (1991a), in investigating a number of astronomy texts, noted that the authors often wrote these in a way that reinforced incorrect mental models. Expressions such as 'the sun went down' reinforce the intuitive mental model that the sun circles the Earth to cause night and day. References in some texts to the tilt of the Earth's axis being responsible for the seasons include such statements as "when the Earth tilts towards the sun it is summer and when it is farther away from the sun, we have winter" (Lucas and Cohen, 1999). In fact, the minor change in distance has nothing to do with the seasons. The tilt changes the incident angle of the sun's radiation, which alters the incident energy per unit area and also the length of daylight. Many of the same conflicts have been observed with tertiary students as with elementary school children in their restructuring of astronomy knowledge (Vosniadou, 1991a, 1991b).

In most instances students have had little or no formal exposure, through previous experiences, to many of the principles and models that explain celestial objects and their motion. These students may retain their naive or intuitive models of the universe with little to no influence from adult scientific models. The pre-Copernican or Ptolemaic belief of an Earth-centered universe would rest comfortably with the intuitive model described by Vosniadou (1991a). Most science textbooks, including those written for undergraduate astronomy students, present students with the scientific model of the solar system, that is, the Copernican model viewed from some point "above." Students are expected to transform their intuitive mental model to the scientific one. This involves complex changes in the frame of reference from an Earth-centered one to a space-centered one, that is, from a viewpoint outside the solar system. In doing this students require a great deal of mental imaging involving spatial visualization and, to some extent, mental rotation.

Students unable to conceptualize these transformations or different viewpoints of the solar system may construct synthetic models to compromise their beliefs in terms of accepted scientific models or views. Vosniadou (1991b) reports that "some students may understand that the Earth rotates on its axis but fail to see how this explains the day/night cycle because they lack a spherical model of the Earth." Alternative models used by students included (a) a disc-Earth mental model and (b) a hollow-Earth mental model. Without an acceptance of a spherical Earth model as a belief, as well as a scientific model, students may have difficulty in describing the nature and causes of the phases of the moon, which is the case with most undergraduates (Vosniadou, 1991a). Studies by Sharp (1996) reported that sixth-grade students were aware of the spherical nature of the Earth and moon and the day/night cycle but had difficulties describing seasons adequately. Most recently, a study by Trumper (2003) found that there were serious discrepancies between pre-service-teacher conceptions and the accepted scientific views, which included day and night, sun's daily apparent motion, solar eclipse mechanisms, and lunar motion, resulting from the influence of students' existing frameworks on their experiences and constructs.

4.4   Teaching and learning strategies

Curricula for the teaching of astronomy should present concepts in an appropriate sequence, and students should be provided with opportunities to examine their personal beliefs and explain their understandings of relevant concepts (Vosniadou, 1991b). A number of researchers have also suggested that the inclusion of activities that develop and enhance student spatial thinking abilities is highly desirable in the design and production of learning materials, and implementation of teaching strategies for astronomy (Broadfoot, 1995; Eylon and Linn, 1988). These activities should be based on direct observation (Lucas and Cohen, 1999; Lucas and Broadfoot, 1991), and spatial training exercises should be included to develop student orientation frameworks and to enhance their spatial visualization of dynamic celestial phenomena (Broadfoot, 1995). Students should also be made aware of their own intuitive beliefs and thought processes through a constructivist approach. Personal or collective investigations by students may be reinforced with a number of observer-centered, concrete three-dimensional models (Bishop, 1990; Broadfoot, 1995; Domenech and Casasus, 1991; Dunlop, 2000).

4.4.1   Visuospatial abilities

In the acquisition and construction of spatial knowledge in astronomy, students often need to imagine how objects look from other viewpoints. To achieve this goal, conceptual restructuring is often necessary, and successful strategies must be planned to ensure that students are trained in the necessary visuospatial skills (Broadfoot, 1995). These strategies could take the form of activities that develop and enhance spatial thinking in the context of astronomy. Design of curricula should also take into account teaching strategies that can achieve modification of the different mental models held by individual students through the mental formation and manipulation of images. Psychological research also indicates a need for spatial development as part of curriculum design for astronomy (Sadler and Luzader, 1990). Students require time and the use of concrete three-dimensional models to develop the complex spatial concepts in astronomy (Bishop, 1990; Sadler and Luzader, 1990).

4.4.2   *Orientation and transformation*

A number of the mental processes may play an important role in shifting students' frame of reference from an egocentric viewpoint to an object-centred one (Heinrich, D'Costa, and Blankenbaker, 1988). Comprehension is dependent on an egocentric orientation used to position items in space with respect to the body, and those perspective shifts can be quite difficult (Piaget and Inhelder, 1956). Many problem-solving exercises in astronomy require students to interpret information from given data and then restructure the information into different viewpoints both verbally and diagrammatically (Vosniadou and Brewer, 1989; Vosniadou, 1991a; Vosniadou, 1991b). Egocentric references are continually needed by students to enable them to navigate the relative positions of celestial objects in space. Therefore students require orientational frameworks for conceptual schema in astronomy which are egocentric but may not be strictly "viewer-centered" or "object-centered."

In studying the sun–Earth-moon system, an object-centered representation is used to describe the orientation and relationship of components. This object-centered viewpoint necessitates considerable mental rotation by the observer from his/her viewpoint as students' perceptions of Earth and space, from a different position, are essentially from a central Earth that also appears "flat." The question that needs to be addressed is how learning materials can be structured to enable students to achieve an outside-looking-in representation of celestial objects from such an inside-looking-out, Earth-centered framework. Therefore, the inclusion of teaching strategies that give training in changing viewpoints with respect to two and three-dimensional models of space and celestial objects is essential.

Broadfoot (1995) proposed that some form of three-dimensional training could contribute significantly to students' understanding of the three-dimensional views and the dynamics of celestial motion. Progressive and continuous mental rotation (transformations) may be used to gain an understanding of the motion of the stars and constellations in the night sky. Field exercises may include those that provide the opportunities for students to mentally change their frame of reference. During field assignments students require initial landmarks as anchors in locating and acquiring knowledge about the groups of stars or constellations (Evans, Marrero, and Butler, 1981). The idea of relating objects to each other rather than through the relationship to the whole was observed in the student use of "pathfinder" strategies that provided a transitional mechanism in locating and monitoring the positions of changing stars and constellations (Broadfoot, 1995). Therefore, the use of pathfinders – readily recognizable stars with imaginary lines joining them – may act as landmarks that students are able to use to navigate their way around the night sky.

The recognition of different views when using telescopes presents further difficulties to students in trying to compare and recognize features on the moon's surface. Quite often the image viewed through a telescope is inverted and flipped, both horizontally and vertically, and students have to deal with extremes of orientation. There is a need to use lunar landmarks as a strategy to recognize the orientation of the moon's virtual (telescopic) image when comparing it to the real or naked-eye view. These necessary shifts in the frames of reference, viewpoints, use of landmarks and rotation have implications for the development of teaching strategies where astronomical perceptions may create possible conflicts of object-centered views with observer-centered views.

To overcome these conceptual difficulties, Lucas and Cohen (1999, p. 15) recommended that "knowledge based on direct observation constitutes an essential foundation for students." To exploit and understand students' prior conceptions, Lucas and Broadfoot (1991) provided a possible observation program, over a year, in which students could challenge their prior beliefs through observational experiences. Taylor (1998) has recommended stages of focus, challenge, application, and reflection as teaching strategies to challenge students in their learning in astronomy.

4.4.3   Use of models

The problem of knowledge representation is central to cognitive stimulation and development and hence good problem-solving skills. In any teaching strategy, it is essential that the teacher or instructor be able to synthesize an accurate picture or model of a student's misconceptions from meager evidence provided through student errors (Good, 1987). The extensive use of models, both static and dynamic, to supplement classroom and field activities, was found to improve students' visuospatial abilities (Broadfoot, 1995). Classroom activities included sketching of astronomical phenomena from text and verbal descriptions, as well as interpretation of observations with reference to accepted scientific views.

Active involvement by students in modeling has been shown to be useful in astronomy teaching (Sadler and Luzader, 1990). Modeling of celestial spheres and star positions in conjunction with real sky activities and events would further enhance students' abilities to convert abstractions and observations to meaningful concepts. Phases of the moon should form part of this introduction to modeling of celestial concepts and natural phenomena that are easily observable. Some researchers (Sweitzer, 1990) even advocate the use of pictorial models, common language analogies and the inclusion of humanistic perspectives. Taylor and Barker (2000) have recommended that mental-model building should form the basis of concept development when studying the sun–Earth–moon system, a conclusion supported by Dunlop (2000), who studied students' learning through engagement in planetarium sessions using three-dimensional planetary models. Students are able to see the dynamic relationships directly; however, concept development needs to be well structured to use models such as orreries, because of the problems of juxtaposition and changes in frames of reference.

4.4.4   Computer modeling

Computer modeling of celestial phenomena has been investigated by some researchers (Mazzolini and Halls, 2000). The use of computer technology in the modeling of astronomical concepts needs to be investigated to establish whether there is any real improvement in the basic understanding of astronomy concepts. There may be a risk of creating a "busy," albeit interesting, curriculum without the sequential scaffolding and development of concepts in astronomy.

4.4.5   Analogies

Vosniadou and Ortony (1989) discuss a number of strategies in the use of analogies to enable student structuring of mental models of celestial phenomena. Between-domain analogies provide the use of metaphors in comparing source and target. For example, the sun can be likened to the nucleus of an atom while the planets can be likened to orbiting electrons. Within-domain analogies have been used to provide causal explanations of day/night cycles and the changing appearance of the moon. It is important to use the Earth as a source analogy, in that students have phenomenal or intuitive experiences about the Earth and the moon (Vosniadou and Brewer, 1987). Taylor (1998) has described the use of an orrery, to illustrate structural relationships between the sun, Earth, and moon, as an analogy. This may have applications in the transformation processes used by students to change their frame of reference.

4.4.6   *Identifying similarities and common identities*

Identifying similarities in students' concepts or beliefs may also be used as a successful strategy (Vosniadou and Ortony, 1989; Glynn, Yeany and Britton, 1991; Gentner and Stevens, 1983). In a holistic view, concepts or percepts may have a dimension that is common on either a global or a dimensional basis. The holistic perception of the moon and sun is that they both "give off light" and are both "round." Comparisons of sizes can only be made by sensory perception, that is, the moon appears larger or about the same size as the sun, when it is in fact much smaller. This real size difference is not readily available or perceivable to students.

The authors propose that, by identifying a common similar element, it should be possible for the teacher to use that sameness with scientific truth as a core concept to bring about a change process in constructing knowledge. In the case of the sun and moon, the core concept would be "round." By recognizing common identities it should be possible for students to use these commonalties as a basis to establish links with the intuitive model of the Earth. This type of strategy forms part of a truly constructivist approach to teaching difficult astronomy concepts that are usually firmly infiltrated with prior beliefs. As a further example of the use of similarities and analogies, a clock could be likened to the heliocentric solar system in which the source is the clock and the solar system is the target. Similarities exist in that the hands of the clock represent time passing, as does the motion of the planets around the sun. This analogy is problematic in that the hands of a clock have fixed motions whereas the planets orbit the sun at different rates with planets closer to the sun completing orbits faster than the planets further away. Therefore, we need to exercise caution in choosing these kinds of analogies as teaching strategies.

4.4.7   *Conceptual restructuring*

Constructivist approaches to the teaching of astronomy have been well described (Bishop, 1990; Domenech and Casasus, 1991; Hill, 1990; Sadler and Luzader, 1990; Verschuur, 1990; Vosniadou, 1991b; Vosniadou and Ortony, 1989). Vosniadou (1991b) has recommended (p. 221) that the design of curricula for the teaching of astronomy should:

(1) present the concepts that comprise a given domain in a sequence consistent with the order in which these concepts are acquired;
(2) create circumstances for students to question their beliefs; and
(3) provide clear explanations of scientific concepts.

However, naive student observations should be used with caution as these may not lead to intended concept development (Broadfoot, 1995). For example, the sphericity of the Earth and the heliocentric nature of the solar system cannot easily be verified by direct observation (Hill, 1990) but are often based on naive theories (Sadler and Luzader, 1990) or rote learning from previous studies.

Predictions should be based on personal or collective investigations by students and reinforced with a number of observer-centered three-dimensional models (Dunlop, 2000; Domenech and Casasus, 1991; Bishop, 1990). These models need to be pertinent and make interactions transparent to students (Hill, 1990). Quite often direct verification of astronomical concepts is not possible from naive observations. The sphericity of the Earth or the heliocentric nature of the solar system are examples.

4.4.8   Historical reconstruction of knowledge

Several researchers (Feteris and Hutton, 2000; Noble, 1999; Verschuur, 1990; Saltiel and Viennot, 1985) have recommended strategies with a strong emphasis on the use of the discovery method to achieve changes in student viewpoints and intuitive knowledge of astronomical phenomena. Saltiel and Viennot propose that "the processes of development of science knowledge in history (on the one hand) and science learning by our students (on the other hand) shows striking analogies." (p. 203) Noble (1999) also showed that exposure of Year 3 students to historical geocentric and heliocentric approaches assisted students in developing a sidereal framework. A number of researchers (Noble, 1999; Broadfoot, 1995; Saltiel and Viennot, 1985) encapsulate a constructivist approach to astronomy teaching based on the historical discoveries and the development of science knowledge in history. Accordingly, a pedagogical approach that provides opportunity for students to construct concepts in astronomy from first-hand observations is recommended. Students need to be given the opportunity to rediscover concepts and relationships from direct observation. This approach is also well supported by Feteris and Hutton (2000) through their approach to teaching astronomy, which involves students in "trivial" cut-and-paste activities that provide students with simple tools to carry out first-hand observations.

4.4.9   Summary

Based on the findings of many researchers, the inclusion of activities that develop and enhance student spatial abilities is highly desirable in the development of learning materials and teaching strategies for astronomy. These activities should be contextual in that spatial training exercises should directly relate to the development of student viewpoints and frameworks as well as to their interpretation of dynamic celestial phenomena. As successful strategies are dependent on the spatial abilities or cognitive resources of individual students, spatial processing strategies should be appropriate to the individual. In taking note of the research, the need for repeated mental shifts should be avoided. In particular, when studying the solar system, the learner must be able to accommodate shifts in mental frameworks alternating between egocentric, heliocentric, and galactic views of the solar system. However, the strategy of using progressive and continuous mental rotation may be used to gain an understanding of the motion of the stars and constellations in the night sky. This is best achieved through a well-sequenced program of practical viewing activities conducted over an extended period of time, which enables students to gain first-hand knowledge about celestial motion. In this way students' own observations would form the bases for restructuring their existing knowledge. Without a well-established spatial understanding of the universe, students will be unable to question entrenched beliefs and evaluate empirical evidence.

Intuitive information can also counter well-designed curricula because of conflict with scientific or observed views. We should, therefore, create circumstances for students to question their beliefs and, at the same time, ensure that we provide clear explanations of scientific concepts. Knowledge acquisition processes should also involve the construction of initial mental models, based on everyday experience, which may be modified to become consistent with scientific models. In designing curricula that cater to the constructivist theory of learning in astronomy we need to give careful thought to the order of the teaching of concepts.

Closely related and causal concepts need to be considered in the design of teaching sequences. For example, an understanding of the concept of light and shadows as well as the dynamic model of the Earth–moon–sun system is a precursor to the development of the correct spherical mental model of lunar phases, which incorporates the changing position of the moon.

The extensive use of conceptual models and analogies would necessarily form an integral part of the construction of students' knowledge and understanding of the complex nature of our dynamic universe. Teachers should provide students with the opportunity to question their entrenched beliefs through long-term periodic observations of celestial phenomena and then use this empirical knowledge to explain and demonstrate the advantage of a new conceptual model. It is only through such an investigative approach, initially focusing on "primitive" intuitive mental models, that students' existing schema can be modified or corrected. It is important to identify intuitive beliefs, challenge student beliefs, and offer different explanations and frameworks to enable students to assimilate and accept scientific beliefs. Diagrams used to illustrate these concepts also need to be clear and logical without a mix of different viewpoints. Students should be provided with the opportunity to rediscover knowledge and understanding about celestial phenomena for themselves in a sequence that follows historical discoveries.

4.5   Recommendations for research

This paper has drawn attention to two important learning and teaching issues:

(1) the difficulties learners face when grappling with and seeking to explain the complexities of astronomical phenomena; and

(2) the approaches and strategies that may be used to enhance the teaching and learning of astronomy.

Based on this background of research findings, the paper concludes with exemplars of possible further research investigations that should be conducted in order to evaluate the effectiveness of these approaches and strategies in overcoming the difficulties experienced by students. Further, it is only through a holistic approach to the analysis of the problems in the teaching and learning of astronomy that gains will be made in this area. Our exemplars of possible research investigations follow.

Learning and teaching
- What thinking processes, with respect to orientation frameworks, do students use when confronting celestial motion problems?
- How do students interpret information from given data and restructure the information into different viewpoints both verbally and diagrammatically?
- What concepts are prerequisites and interdependent in astronomy learnings?

Teacher training
- What is the exact nature of, and what factors affect, the designed and implemented curricula in astronomy in our tertiary institutions?
- Do these curricula produce teachers who are competent in teaching astronomy?

Primary and secondary schools

- What is the exact nature of, and what factors affect, the designed and implemented curricula in astronomy in our primary and secondary schools?
- Do these curricula engage students in challenging and constructing their knowledge and understanding of astronomy?

References

Astronomical Society of Australia 2002, University astronomy courses in Australia. URL: http://msowww.anu.edu.au/~pfrancis/asa/astro_courses.htm

Bishop, J. E. 1990, "Dynamic human (astronomical) models," in J. M. Pasachoff and J. R. Percy, eds., *The Teaching of Astronomy*, IAU Colloquium 105, Cambridge: Cambridge University Press.

Broadfoot, J. M. 1995, "Development of visuospatial abilities among undergraduate astronomy students," Unpublished Masters thesis. Curtin University of Technology, Perth.

Brooks, G. P. 1987, "Contributions to the history of psychology: XLII. Note on Isaac Watts, astronomy, and mental models and moral improvement," *Psychological Reports*, **60**, 740–2.

Domenech, A. and Casasus, E. 1991, "Galactic structure: a constructivist approach to teaching astronomy," *School Science Review*, **72**(260), 87–93.

Dunlop, J. 2000, "How children observe the universe," *Publications of the Astronomical Society of Australia*, **17**(2), 194.

Ekstrom, R., French, J., Harmon, H., and Derman, D. 1976, "Manual for kit of factor-referenced cognitive tests," Princeton, NJ: Educational Testing Service.

Evans, G. W., Marrero, D. G., and Butler, P. A. 1981, "Environmental learning and cognitive mapping," *Environment and Behaviour*, **13**(1), 83–104.

Eylon, B. and Linn, M. C. 1988, "Learning and instruction: an examination of four research perspectives in science education," *Review of Educational Research*, **58**(3), 251–301.

Feteris, S. and Hutton, D. 2000, "Astronomy laboratory: what are we going to make today?" *Publications of the Astronomical Society of Australia*, **17**(2), 116.

Finegold, M. and Pundak, D. 1990, "Students' conceptual frameworks in astronomy," *Australian Science Teachers Journal*, **36**, 76.

Gentner, D. and Stevens, A. L., eds. 1983, *Mental Models*, Hillsdale, NJ: Erlbaum.

Glynn, S. M., Yeany R. H., and Britton B. K. 1991, "A constructive view of learning science," in S. M. Glynn, R. H. Yeany, and B. K. Britton, eds., *The Psychology of Learning Science*, Hillsdale, NJ: Erlbaum, 3–19.

Good, R. 1987, "Artificial intelligence and science education," *Journal of Research in Science Teaching*, **24**(4), 325–42.

Heinrich, V., D'Costa, A., and Blankenbaker, K. 1988, "A spatial visualisation test for selecting engineering students. Technical Report No. 143," for the US Department of Education, Ohio State University.

Hill, L. C. (Jr). 1990, "Spatial thinking and learning astronomy: The implicit visual grammar of astronomical paradigms," in J. M. Pasachoff and J. R. Percy, eds., *The Teaching of Astronomy*, IAU Colloquium 105, Cambridge: Cambridge University Press.

Hintzman, D. L., O'Dell, C. S., and Arndt, D. R. 1981, "Orientation in cognitive maps," *Cognitive Psychology*, **13**, 149–206.

Howard, I. P. 1982, *Human Visual Orientation*, New York: Wiley.

Humphreys, G. W. 1983, "Reference frames and shape perception," *Cognitive Psychology*, **15**, 151–96.

Just, M. A. and Carpenter, P. A. 1985, "Cognitive coordinate systems: Accounts of mental rotation and individual differences in spatial ability," *Psychological Review*, **92**, 137–71.

Kosslyn, S. M., Cave, C. B., Provost, D. A., and von Gierke, S. M. 1988, "Sequential processes in iImage generation," *Cognitive Psychology*, **20**, 319–43.

Lucas, K. B. and Broadfoot, J. M. 1991, "Astronomical activities throughout the year for primary school students," Paper presented at the Annual Conference of the Science Teachers' Association of Queensland, Kelvin Grove, Queensland.

Lucas, K. B., and Cohen, M. R. 1999, "The changing seasons: teaching for understanding," *Australian Science Teachers' Journal*, **45**(4), 9–17.

Marr, D. 1982, *Vision*, San Francisco: Freeman.

Mazzolini M. and Halls, B. 2000, "*Astro concepts* – learning underlying physics principles in conceptual astronomy," *Publications of the Astronomical Society of Australia*, **17**(2), 149.

Noble, A. 1999, "Using prehistory of science to teach astronomy in the primary school," Paper presented at the annual meeting of the Australasian Science Education Research Association, Rotorua, NZ.

Piaget, J. and Inhelder, B. 1956, "The child's conception of space," English translation: Langdon, F. J. and Lunzer, J. L. 1967, London: Compton.

Pinker, S. and Finke, R. 1980, "Emergent two-dimensional patterns in images rotated in depth," *Journal of Experimental Psychology: General*, **109**, 354–71.

Rock, I., Wheeler, D., and Tudor, L. 1989, "Can we imagine how objects look from other viewpoints?" *Cognitive Psychology*, **21**, 185–210.

Sadler, P. M. and Luzader, W. M. 1990, "Science teaching through its astronomical roots," in J. M. Pasachoff and J. R. Percy, eds., *The Teaching of Astronomy*, IAU Colloquium 105, Cambridge: Cambridge University Press.

Saltiel, E. and Viennot, L. 1985, "What do we learn from similarities between historical ideas and the spontaneous reasoning of students?" in P. J. Lijnse, ed., *The Many Faces of Teaching and Learning Mechanics: Conference on Physics Education*, Utrecht, Netherlands: GIREP/SVO/UNESCO, 199–214.

Sharp, J. 1996, "Children's astronomical beliefs: A preliminary study of year 6 children in south-west England," *International Journal of Science Education*, **18**, 685–712.

Shepard, R. N. and Cooper, L. A. 1982, *Mental Images and their Transformation*, Cambridge, MA: Bradford/MIT Press.

Stannard, P. and Williamson, K. 1999, *Science World* (Four Book Series), Melbourne: Macmillan.

Sweitzer, J. S. 1990, "Strategies for presenting astronomy to the public," in J. M. Pasachoff and J. R. Percy, eds., *The Teaching of Astronomy, IAU Colloquium 105*, The Proceedings of the 105th Colloquium of the International Astronomical Union, Williamstown, Massachusetts, July 26–30, 1988, Cambridge: Cambridge University Press.

Takano, Y. 1989, "Perception of rotated forms: A theory of information types," *Cognitive Psychology*, **21**(1), 1–59.

Taylor, I. J. 1998, "Children's mental models: How do they construct them?" URL: http://www2.deakin.edu.au/faculty/Math_Sci_Enviro/Conferences/Papers_1998/Taylor.pdf

Taylor, I. J. and Barker, M. 2000, "Forgettable facts or memorable models? Arguments for a mental model building approach to astronomy education," Paper presented at the annual meeting of the Australasian Science Education Research Association, Fremantle, WA.

Thickett, G., Stamell, J., and Thickett, L. 2000, *Science Tracks* (Four Book Series), Melbourne: Macmillan.

Treagust, D. F. and Smith, C. L. 1989, "Secondary students' understanding of gravity and the motion of planets," *School Science and Mathematics*, **89**(5), 380–91.

Trumper, R. 2003, "The need for change in elementary school teacher training – a cross-college age study of future teachers' conceptions of basic astronomy concepts," *Teaching and Teacher Education*, **19**(3), 309–23.

Verschuur, G. L. 1990, "The thrill of discovery," in J. M. Pasachoff and J. R. Percy, eds., *The Teaching of Astronomy, IAU Colloquium 105*, The Proceedings of the 105th Colloquium of the International Astronomical Union, Williamstown, Massachusetts, July 26–30, 1988, Cambridge: Cambridge University Press.

Vosniadou, S. 1989, "Knowledge acquisition in observational astronomy," Paper presented at Annual Meeting of the American Educational Research Association. Washington, DC.

Vosniadou, S. 1991a, "Conceptual development in astronomy," in S. M. Glynn, R. H. Yeany, and B. K. Britton, eds., *The Psychology of Learning Science*, Hillsdale, NJ: Erlbaum, 149–77.

Vosniadou, S. 1991b, "Designing curricula for conceptual restructuring: Lessons from the study of knowledge acquisition in astronomy," *Journal of Curriculum Studies*, **23**(3), 219–37.

Vosniadou, S. and Brewer, W. F. 1987, "Theories of knowledge restructuring in development," *Review of Educational Research*, **57**(1), 51–67.

Vosniadou, S. and Brewer, W. F. 1989, "The concept of the Earth's shape: A study of conceptual change in childhood," Technical Report No. 467. Office of Educational Research and Improvement (ED), Washington, DC.

Vosniadou, S. and Ortony, A. 1989, "Similarity and analogical reasoning: a synthesis," in S. Vosniadou and A. Ortony, eds. *Similarity and Analogical Reasoning*, Cambridge: Cambridge University Press, 1–17.

Yackel, E. and Wheatley, G. H. 1990, "Promoting visual imagery in young pupils," *Arithmetic Teacher*, **37**(6), 52–8.

Comments

Bill MacIntyre: In your survey of tertiary institutes, were there specific astronomy education courses/subjects presented for pre-service teacher trainees?

John Broadfoot: There are not any specific teacher education courses that pertain to astronomy education.

5

A contemporary review of K–16 astronomy education research

Janelle M. Bailey and Timothy F. Slater
Conceptual Astronomy and Physics Education Research (CAPER) Team,
Department of Astronomy, University of Arizona

Abstract: Despite astronomy's widespread inclusion in curricula prior to the twentieth century, educational research in astronomy is a relatively new endeavor. As the field of astronomy education research grows, many may find it useful to know what has been done so far. Starting with and expanding beyond the SABER database, a systematic review and classification of the K–12 and higher education literature was performed. Some of the research themes that emerged include: student beliefs and misconceptions; collaborative learning; the large lecture classroom; and education in planetariums. Key studies in these areas are described and a bibliography is presented.

Astronomy education research (AER) uses the systematic techniques honed in science education and physics education research to understand what and how students learn about astronomy, and determine how instructors can create more productive learning environments for their students. As the field of AER grows – and it is doing so vigorously – many readers may find it useful to have a concise summary of what has been published to date. The purpose of this paper is to summarize and categorize the various research projects in astronomy education in order to set the stage for subsequent efforts. In researching the field, a number of information sources were consulted. Three electronic resources served as the starting point for the review: the SABER (Searchable Annotated Bibliography of Education Research) database, the American Astronomical Society's education bibliography (http://www.aas.org/education/biblio_list.html), and the Astronomical Society of the Pacific's education bibliography (Fraknoi, 1998). Further references were found in a variety of journals, often through a "snowballing" technique of looking through an article's references for new articles. Journals included (but are not limited to) *Astronomy Education Review, Journal of Research in Science Teaching, Journal of College Science Teaching, Science Education*, and *The Physics Teacher*.

In order to limit the extent of this diverse field, we have made purposeful choices about the references included here. In this paper, research is defined as those studies that attempt to systematically analyze issues such as student conceptions on a topic or the effectiveness of a particular instructional or curricular intervention. An analysis of the literature revealed three main research areas: student understanding, instructional methods, and teacher understanding. Outside of this definition, and the scope of this paper, lie a multitude of additional works, including descriptions of innovative activities, curricula, or instructional techniques; information regarding the US National Research Council's "National Science Education Standards" (1996, hereafter NSES); and articles on the teaching of the nature of science. A more comprehensive review of the literature can be found in Bailey and Slater (2003b).

5.1   Research on student understanding

Within the research on student understanding are a number of topics. The most thoroughly investigated area is the shape of Earth (including notions of gravity) with students in grades 2–8 (approximate ages 7–13) in the USA, Nepal, Israel, and the UK. Understanding of lunar phases has been investigated for students in grades 3–5 and university, as well as pre-service and in-service teachers in the USA. Seasons have been investigated with university students and pre- and in-service teachers in the USA. Diurnal motion (day/night cycle) is a topic studied with students in grades 1–5 and pre-service teachers in the USA. American university, high-school, and middle-school students have been asked about their understanding of astrobiology and cosmology issues. Finally, the Astronomy Diagnostic Test (ADT) has been used with university students across the United States.

5.2   An example: shape of Earth and related issues

Joscph Nussbaum and his colleagues conducted some of the earliest studies into student conceptual understanding in earth and space sciences. Nussbaum and Novak (1976) used semi-structured clinical interviews to uncover student conceptions about the shape of Earth, its position in space, and how gravity affects falling objects. From interviews with US second-grade students, the researchers discovered five recurring "notions" describing the shape of Earth (only one of which is scientifically accurate), despite many student comments that at the surface appear to be correct. These notions included (a) the flat but round (like a pancake) Earth, with no concept of "space"; (b) a spherical Earth but with an external ground that "limits" things that fall, defining an absolute down (as if the ball-shaped Earth were separate from the everyday ground the student experiences); (c) a spherical Earth with some idea of unlimited space but maintaining an absolute down concept (where things on the "bottom" of Earth might fall off); (d) a spherical Earth with space around it, but persistent problems with falling objects; and (e) a spherical Earth, surrounded by space, with gravity pulling objects toward the center of the planet (scientifically accurate). Illustrations of these notions can be seen in Nussbaum and Novak.

Nussbaum (1979) later expanded this work in an attempt to generalize the earlier results on Earth-shape notions. The interview format was adapted into a multiple-choice instrument with additional space to allow students to provide explanations or drawings; the children were also given three-dimensional models to manipulate during their explanations. A total of 240 Israeli students in grades 4–8 were interviewed and their notions categorized. The number of children with scientifically accurate models increased somewhat with age, providing indirect support for the idea that a developmental trend exists. Using the same methods as the Nussbaum and Novak (1976) study, Mali and Howe (1979) looked at the cognitive development of Nepali children. Somewhat surprisingly, they also found all five of the Nussbaum and Novak notions to be present in this population, which is very different from US children, leading to the idea that the notions are not culturally engendered. Mali and Howe also found a slight progression toward more advanced notions with age, although the notion/age relationship tended to be systematically shifted relative to the results of the American studies. Nussbaum and Sharoni-Dagan (1983) further addressed alternative conceptions in a study of the effectiveness of audio-tutorials, where, contrary to popular belief, they found that younger children were able to seemingly grasp the abstract notions of Earth as a cosmic body when presented with appropriately targeted instruction.

Klein (1982) addressed similar issues of student understanding in a cultural comparison of Mexican-American and Anglo-American second-grade students. She found that students held a number of ideas contrary to concepts they had been exposed to in the classroom, and that, like the students in previously described studies, they had little understanding of Earth as a ball in space. Sneider and Pulos (1983) extended the Nussbaum (1979) study to California students in grades 3–8, revamping the notions into two separate categories of shape and gravity. They found a strong age-related trend as before, and through multiple regression analysis found that differences in verbal ability and spatial reasoning, as well as gender, had the most influence on which notion a student holds.

Jones, Lynch, and Reesink (1987) looked at third- and sixth-grade students' beliefs about Earth's shape, size, and location through a combined use of interviews and model demonstrations. Like those before it, this study found that a larger portion of older students demonstrated a scientifically correct understanding of these topics than younger students. Baxter (1989) had similar findings in England and used this result to argue that student learning reflects the historical development of astronomy: "The reference to historical ideas possibly makes pupils feel more comfortable when they realize that their notions, although incorrect in the light of scientific advancement, were once the popular view" (Baxter, 1989, p. 512). Stella Vosniadou and her colleagues explored children's understandings by using a cognitive science perspective. Vosniadou and Brewer (1992) criticized previous researchers for not making sufficiently explicit their notion or category identification criteria and for providing little or no information on the child's consistency in using that notion in seemingly diverse scenarios. Results from their research into children's understanding of the shape of Earth were consistent with prior work as described above.

At this point the reader might wonder why this degree of attention has been devoted to these ideas. First, it reflects the community's excitement over these unexpected results. Second, the researchers had an intuitive feeling that a student's conception of gravity, a notoriously difficult area of teaching and learning, was dependent upon an accurate conception of a spherical Earth. An example of where this difficulty might make itself known is in the cognitive conflict a student encounters when trying to understand why penguins in the Antarctic (on the "bottom" of Earth, from the point of view of students in the Northern Hemisphere) do not fall off. In the end, this research has not yet dramatically impacted instruction.

5.3   Examples of studies about lunar phases

A common result in studies of student understanding about lunar phases is the misconception known as the "eclipse model," where students believe the phases of the moon are caused by Earth's shadow falling on the moon, thus blocking its light. Stahly, Krockover, and Shepardson (1999) interviewed four US third-grade students. Additional data were collected in the form of written tasks, classroom observations, and teacher feedback. Comparisons between pre- and post-instruction interviews demonstrated that the students showed a positive conceptual change over an instruction intervention characterized as a three-week, multiple-component lesson using three-dimensional models, with post-instructional explanations showing more details and more scientifically accurate ideas.

For older students, Lindell (2001) developed the Lunar Phases Concept Inventory (LPCI) from interviews with undergraduate students to measure conceptual change in their understanding of the moon's phases. Her pretest results confirmed earlier work with younger students suggesting that scientifically inaccurate ideas persist into college. The LPCI, a multiple-choice instrument that can be quantitatively analyzed, was used to evaluate the effectiveness of an in-class group activity designed to address these concepts using a constructivist approach.

Trundle, Atwood, and Christopher investigated the understanding of moon phases held by pre-service elementary teachers (Trundle, Atwood, and Christopher, 2002) and in their report provide a table describing the results of many previous studies on this topic. The researchers collected qualitative data through classroom observation, structured interviews, and document analysis. This triangulation of data through multiple sources is a useful and increasingly common way of checking for valid interpretations. A total of 78 participants were interviewed once or twice to determine their mental models. As in most of the earlier studies on this topic, the most common alternative conception about the lunar phases is that they result from eclipses by Earth's shadow. A later report (Trundle, Atwood, and Christopher, 2003) describes the results of additional post-instruction interviews, 6 and 13 months after the instructional period, with a subset of 12 of the participants. At these interview instances, seven still held scientific understanding and two demonstrated scientific fragments. The other three reverted to alternative conceptions. The researchers also specified that two common problems seen in the pre-instruction interviews were that the participants often did not know that the moon orbits Earth, or that the moon is always half illuminated. Without this knowledge, the researchers speculate, students have no reason to question the common shadow-eclipse model. Fanetti (2001) interviewed 50 college students and administered an open-ended survey to more than 700 students to investigate their understanding of lunar phases. Her main conclusion from these data is that a lack of understanding of the moon–Earth system's scale is the main reason for student difficulties in understanding the phases of the moon.

5.4   An example: Student understanding of cosmology

In an effort to understand students' preexisting mental models about cosmology that might be poised to interfere with instruction, Prather, Slater, and Offerdahl (2002) report preliminary results from a survey of nearly 1,000 middle school, high school, and college students. On a student-supplied written-response survey, students were asked to describe and draw their understanding of the Big Bang. When the responses were inductively analyzed into themes, the authors found that 62 per cent of middle school, 70 per cent of high school, and 80 per cent of college students believed the Big Bang to be an explosion that organized preexisting matter. Between 25 per cent and 37 per cent of each group said the Big Bang describes the creation of planetary systems.

5.5   Instructional effectiveness research

The area of instructional effectiveness research is not as extensive as that of research on student understanding, but some themes are present. Use of the Personalized System of Instruction (often called the Keller Plan) in introductory astronomy courses at the university level has been investigated. Instructors have reported on the development and implementation of conceptual university courses, as well as the use of collaborative groups in introductory courses. Great Explorations in Math and Science (GEMS, created by the Lawrence Hall of Science) units relating to astronomy are used in some US K–8 classrooms. Finally, investigations into the efficacy of technology in K–16 astronomy education, including distance learning and planetariums, have been reported.

5.6   Instructional effectiveness research: lunar phases

Callison and Wright (1993) investigated the use of different three-dimensional models to teach about lunar phases, focusing especially on the interaction of the students' spatial ability and reasoning levels in their ability to develop mental models. Barnett and Morran (2002) investigated the use of project-based curriculum from the Challenger Center (http://www.challenger.org) in a fifth-grade classroom to teach about the phases of the moon and eclipses. They incorporated findings from research into student understanding by the sequencing of projects, with the report focusing only on the final two projects (covering the relative Earth–moon–sun positions and eclipses). The researchers found that their students were able to increase their understanding of moon phases (as defined by having a more scientifically complete understanding) after having experienced this series of lessons. Many of the studies described earlier (Lindell, 2001; Stahly, Krockover, and Shepardson, 1999; Trundle, Atwood, and Christopher, 2002, 2003) also describe a particular instructional intervention in relation to student understanding of lunar phases.

Abell and colleagues used a six-week-long moon observation project in their elementary science teaching methods course to emphasize aspects of the nature of science as well as content to pre-service teachers (Abell, Martini, and George, 2001). The details of the goals, methods, and assessments used in this process are described in a later report (Abell, George, and Martini, 2002). Students recorded daily moon observations in journals, eventually moving to identifying patterns, making predictions, and offering explanations in the journals as well. While students were able to recognize the importance of observation in science, they were not often able to articulate the different roles that observation can play. Finally, both small and large group discussions were used to emphasize the social nature of science. Most of the students recognized the importance of social collaboration in their own learning process, but failed to make the connection between this aspect and the work of scientists.

5.7   Research on teacher understanding

The research base on teacher understanding in astronomy is very small to date, although this is somewhat reflective of general education research as well. Researchers have used pre-service and in-service teachers as the participants in investigations of understanding (e.g., Atwood and Atwood, 1996 on seasons; or Trundle *et al.*, 2002, 2003 on lunar phases) and have looked at differences between "expert" and "novice" teachers (Barba and Rubba, 1992). Lightman and Sadler (1988, 1993) investigated whether teachers could accurately predict student performance on a multiple-choice test given both before and after an astronomy course. Slater (1993) used a constructivist approach in the development and evaluation of an in-service course in astronomy for elementary and middle-schoolteachers. Ashcraft and Courson (2003) investigated the effectiveness of an inquiry activity with pre-service teachers.

5.8   Descriptions of curricula and other materials

A wide body of literature exists describing individual curriculum pieces or activities in astronomy. Topics include, for example, size and scale of various astronomical objects; observing projects on the motions of the sun, moon, stars, or planets; comparative planetology; using real astronomical data via the Internet; and astrology. There are also descriptions of outreach and informal education programs. The Astronomical Society of the Pacific's webpage (Fraknoi, 1998) contains an extensive list of curriculum materials.

5.9　Conclusions

As the field of AER is reviewed, a few surprises can be seen. Although the research base on student understanding is the largest subtopic, it still actually contains very few topics investigated to any depth (especially when compared with physics education research, for example). The existing bibliographies and databases are largely incomplete, perhaps in part because of AER's rapid growth in recent years. There are very few systematic studies of the effectiveness of new curriculum materials or instructional strategies. Perhaps the largest piece of AER that is missing still today is the practice–theory connection that makes use of the theory-driven research in real classrooms. Another area of research that has yet to be aggressively explored is the underlying cause of student difficulties in astronomy. Finally, there is very little existing work on sociocultural issues (such as gender or ethnicity) in AER.

5.10　The future of astronomy education research

As a vibrant young field, astronomy education research holds several possibilities for future directions. Additional research is needed in the area of student understanding, on a wide variety of astronomical topics. Ongoing work by the Conceptual Astronomy and Physics Education Research (CAPER) Team at the University of Arizona includes, for example, studies on student understanding of the properties and evolution of stars (Bailey and Slater, 2003a); the special properties of water, especially as it relates to the origin and evolution of life; and sonification[1] of data from the Mars Gamma Ray Spectrometer. The systematic evaluation of instructional strategies and curriculum materials is another area that may be expanded in the future. As an example, only one of the astronomy-related GEMS guides has been described in the research literature. There are only a handful of studies that look at sociocultural issues in astronomy, and most of these have focused on gender (e.g., Skala, Slater, and Adams, 2000). Some work has begun on astronomy and the nature of science (see e.g., Brickhouse *et al.*, 2000, 2002) but there is much to be done in the future. This wide array of possible research paths makes astronomy education research an exciting and dynamic young field.

References

Abell, S., George, M., and Martini, M. 2002, "The moon investigation: instructional strategies for elementary science methods," *Journal of Science Teacher Education*, **13**(2), 85.

Abell, S., Martini, M., and George, M. 2001, "'That's what scientists have to do': pre-service elementary teachers' conceptions of the nature of science during a moon investigation," *International Journal of Science Education*, **23**(11), 1095.

Ashcraft, P. and Courson, S. 2003 (March), "Effects of an inquiry-based intervention to modify pre-service teachers' understanding of seasons," Paper presented at the Annual Meeting of the National Association for Research in Science Teaching, Philadelphia, PA.

Atwood, R. K. and Atwood, V. A. 1996, "Pre-service elementary teachers' conceptions of the causes of seasons," *Journal of Research in Science Teaching*, **33**, 553.

Bailey, J. M. and Slater, T. F. 2003a, "Initial investigation of students' conceptions of star formation," Paper presented at the Summer Meeting of the American Association of Physics Teachers, Madison, WI.

Bailey, J. M. and Slater, T. F. 2003b, "A review of astronomy education research," *Astronomy Education Review*, **2**(2), 20.

Barba, R. and Rubba, P. A. 1992, "A comparison of pre-service and in-service earth and space science teachers' general mental abilities, content knowledge, and problem-solving skills," *Journal of Research in Science Teaching*, **29**, 1021.

---

[1] The term "sonification" refers to the process of taking data from different channels of the Mars Gamma Ray spectrometer and giving each data point a "sound." This essentially turns the data base into a sort of symphony.

Barnett, M. and Morran, J. 2002, "Addressing children's alternative frameworks of the moon's phases and eclipses," *International Journal of Science Education*, **24**(8), 859.

Baxter, J. 1989, "Children's understanding of familiar astronomical events," *International Journal of Science Education*, **11**, 502.

Brickhouse, N. W., Dagher, Z. R., Letts IV, W. J., and Shipman, H. L. 2000, "Diversity of students' views about evidence, theory, and the interface between science and religion in an astronomy course," *Journal of Research in Science Teaching*, **37**(4), 340.

Brickhouse, N. W., Dagher, Z. R., Shipman, H. L., and Letts IV, W. J. 2002, "Evidence and warrants for belief in a college astronomy course," *Science and Education*, **11**(6), 573.

Callison, P. L. and Wright, E. L. 1993 (April), "The effect of teaching strategies using models on pre-service elementary teachers' conceptions about Earth–sun–moon relationships," Paper presented at the Annual Meeting of the National Association for Research in Science Teaching, Atlanta, GA.

Fanetti, T. M. 2001, "The relationships of scale concepts on college age students' misconceptions about the cause of lunar phases," Unpublished Master's thesis, Iowa State University, Ames.

Fraknoi, A. 1998, "Astronomy education: A selective bibliography," Retrieved April 6, 2003, from http://www.astrosociety.org/education/resources/educ_bib.html

Jones, B. L., Lynch, P. P., and Reesink, C. 1987, "Children's conceptions of the Earth, sun and moon," *International Journal of Science Education*, **9**, 43.

Klein, C. 1982, "Children's concepts of the Earth and the sun: A cross-cultural study," *Science Education*, **65**, 95.

Lightman, A. and Sadler, P. M. 1988, "The Earth is round? Who are you kidding?" *Science and Children*, **25**(5), 24.

Lightman, A. and Sadler, P. M. 1993, "Teacher predictions versus actual student gains," *The Physics Teacher*, **31**, 162.

Lindell, R. S. 2001, "Enhancing college students' understanding of lunar phases," Unpublished doctoral dissertation, University of Nebraska, Lincoln.

Mali, G. and Howe, A. 1979, "A development of Earth and gravity concepts among Nepali children," *Science Education*, **63**, 685.

National Research Council 1996, National Science Education Standards, Washington, DC: National Academy Press.

Nussbaum, J. 1979, "Children's conception of the Earth as a cosmic body: A cross age study," *Science Education*, **63**, 83.

Nussbaum, J. and Novak, J. 1976, "An assessment of children's concepts of the Earth utilizing structured interviews," *Science Education*, **60**, 535.

Nussbaum, J. and Sharoni-Dagan, N. 1983, "Changes in second grade children's preconceptions about the Earth as a cosmic body resulting from a short series of audio-tutorial lessons," *Science Education*, **67**, 99.

Prather, E. E., Slater, T. F., and Offerdahl, E. G. 2002, "Hints of a fundamental misconception in cosmology," *Astronomy Education Review*, **1**, 2. Retrieved on November 7, 2002, from http://aer.noao.edu/AERArticle.php?issue=2&section=2&article=2

Skala, C., Slater, T. F., and Adams, J. P. 2000, "Qualitative analysis of collaborative learning groups in large enrollment introductory astronomy," *Publications of the Astronomical Society of Australia*, **17**, 185.

Slater, T. F. 1993, "The effectiveness of a constructivist epistemological approach to the astronomy education of elementary and middle level in-service teachers," Unpublished doctoral dissertation, University of South Carolina, Columbia.

Sneider, C. I. and Pulos, S. 1983, "Children's cosmographies: Understanding the Earth's shape and gravity, *Science Education*, **67**(2), 205.

Stahly, L. L., Krockover, G. H., and Shepardson, D. P. 1999, "Third grade students' ideas about the lunar phases," *Journal of Research in Science Teaching*, **36**(2), 159.

Trundle, K. C., Atwood, R. K., and Christopher, J. E. 2002, "Preservice elementary teachers' conceptions of moon phases before and after instruction," *Journal of Research in Science Teaching*, **39**(7), 633.

Trundle, K. C., Atwood, R. K., and Christopher, J. E. 2003 (March), "Preservice elementary teachers' conceptions of moon phases: A longitudinal study," Paper presented at the Annual Meeting of the National Association for Research in Science Teaching, Philadelphia, PA.

Vosniadou, S. and Brewer, W. F. 1992, "Mental models of the Earth: A study of conceptual change in childhood," *Cognitive Psychology*, **24**(4), 535.

Comments

Martin George: Planetariums are excellent places to research the astronomy knowledge and ability of both students and teachers. There is a wide variety of such ability. Relevant papers have, over the years, been published as a part of the International Planetarium Society (IPS) conferences, and in the Society's magazine, *The Planetarian*.

John Percy: In North America, there is often very little contact between astronomers, faculties of education, and education researchers. At the University of Toronto, I frequently did workshops for the Faculty of Education, but I was ignored by the education research institute until a graduate student got me in "the back door" and introduced me to her supervisor. Eventually I received a formal cross-appointment.

6

Implementing astronomy education research

Leonarda Fucili

G. G. Belli, Lyceum, and Supervisor in Teacher Development School, "SSIS LAZIO," Rome, Italy

Abstract: As a teacher with a special interest in astronomy, I have experimented for more than ten years with ways in which astronomy might be taught, and used to introduce young students to the complexity of science. My research and teaching are founded on the belief that the effective learning and understanding of astronomical concepts are strongly related to the perception of phenomena, and to the emotions that nature and the sky bring out in us. Students must be guided to find, in the sky, the same fascination and wish for knowledge that has always led mankind to observe astronomical phenomena and organize space and time. Looking at what happens in my class, I will sketch some examples of methods and astronomical activities which create responsive and effective learning environments. I will focus on some competences and conditions for the teaching of basic astronomy in order to encourage the enthusiasm of students, to improve their understanding of science, and their appreciation of its role in making sense of the world. Thus, scientific knowledge becomes culture, and astronomy makes a great contribution, not only in science, but in education.

6.1   Premise

Astronomy plays an essential role in human culture: there is nothing quite like the study of astronomy to capture the imaginations of our students, to make them understand phenomena and introduce them to the fundamental ideas and methods of science and mathematics. But if we want to improve a basic astronomy education and develop a general interest in science, we have not only to give students knowledge and facts, but essentially to guide them to know what to look for, to help them to understand what they know or they are learning, how we have proceeded in our current state of knowledge and how one can proceed to know more. Furthermore, the understanding of astronomical concepts is strongly related to observations of phenomena and to emotions that nature and sky bring out in us. Before students study theory in textbooks, they must be guided to find in the sky the same fascination and wish of knowledge that has always led mankind to observe astronomical phenomena and organize space and time. This is the focus of my research on the effective teaching/learning of astronomy and the goal of my teaching.

6.2   Teaching and learning astronomy
How to improve the students' awareness, understanding, and appreciation of astronomy and not to intensify misconceptions

More than 20 years of teaching, curriculum development and research in astronomy education have made me aware of the conditions of effective teaching of astronomy at the elementary school level, in order to encourage the enthusiasm of students and to increase their familiarity with astronomy, their understanding of science and appreciation of its role

Fig. 6.1. Intensifying misconceptions about orientation.

in making sense of the world. Learning astronomy at the basic level requires a direct involvement, student participation and observations of real phenomena. If we just tell our students "how it is," it doesn't mean they will understand concepts. We have to be patient: students need a long time to observe and to make sense of their observations, to record data and find patterns in them, to build and use their own instruments, to construct models, and gradually to explain the phenomena and to acquire the fundamental concepts of astronomy.

The importance of outdoor observations for the teaching of astronomy has to be emphasized: they lead students to ask questions and improve their understanding. However, students have to be guided in "knowing what to look for," because usually they don't observe much more than things they are already familiar with. They will discover that a surprising abundance of observations is possible and they will find connections between what they learn at school and the real world. On the other hand, the usual school practice, with short-term activities and lessons inside the classroom, following a fixed curriculum and textbooks, sometimes creates a negative impression of astronomy and intensifies misconceptions.

It can be interesting to examine some examples of "mistakes" in ordinary illustrations in textbooks for primary school. One simple example refers to orientation (Fig. 6.1). The drawing doesn't connect orientation with the daily sun motion and it intensifies the misconception that south is down and north is up, according to the geographical maps hanging on the school walls.

The second example concerns the phases of the moon. They are often explained in textbooks with a quite complex diagram of the Earth and moon where two different points of view are shown: the moon phases as seen from space, and seen side-on from the Earth. Such a drawing is too abstract for young students to understand, but things become much more difficult when the drawing is wrong: in Fig. 6.2, the sun is pictured close to the moon and nearly of the

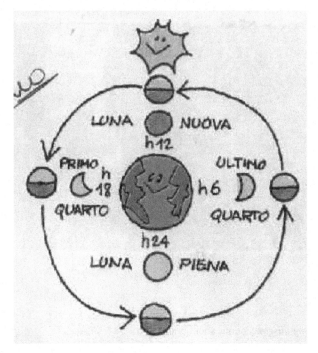

Fig. 6.2. The moon phases.

same size, so it is not understandable how it illuminates half of the moon's sphere, producing the phases. Secondly, there is a mistake concerning the first and the last quarter, how they would appear to a person standing on the Earth: they have to be exchanged. In this way, misconceptions remain, scientific concepts are not developed, and astronomy becomes a difficult subject: it is a reason why few teachers include astronomy in their curriculum and present it effectively.

6.3   Ten competences for teaching astronomy

Looking at what happens in my class, I shall first explain the context: I teach in "Scuola Media," usually the same class for three years, from grade 6 to 8. Students are of all abilities; aged 11–14. I teach mathematics and science to each class for six hours a week, and an integrated project for two hours. This gives me the opportunity to develop long-term astronomical activities and lessons, as well as the mathematical knowledge, skills and concepts students need for the understanding of astronomy.

As an integrated project, I developed an interdisciplinary introductory astronomy course. The curriculum is flexible, modular and in-progress, an open path focused on concepts. Knowledge construction is never the repetition of a fixed path, so my guideline is neither the book nor a fixed and linear sequence of subjects.

The essentials of the course are organized in "concept maps" that focus the essentials of the discipline and the connections between goals, topics and concepts. They can be surfed in different directions, taking into account students' pre-knowledge, obstacles in learning, interests, and abilities. The map of goals (Fig. 6.3) and the map of subjects (Fig. 6.4) are

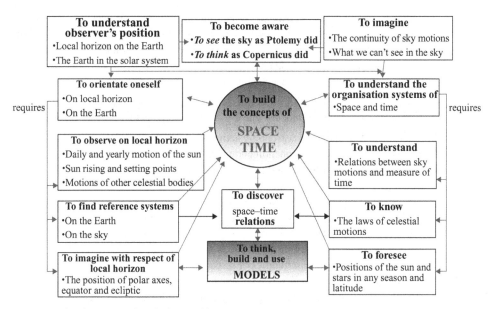

Fig. 6.3. Map of goals for teaching astronomy.

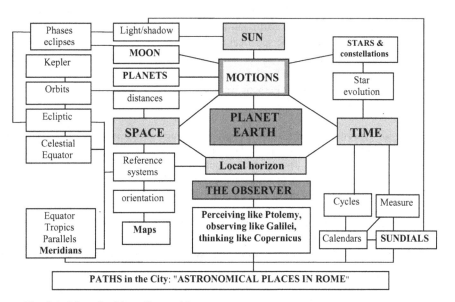

Fig. 6.4. Map of subjects for teaching astronomy.

shown here. I'll sketch some examples of methods and activities that I have taken out of my research and practice in teaching astronomy which I consider essential in order to create responsive learning environments and to reach effective learning of astronomical concepts.

(1) **Inducing students to take an interest in the content**. To see something meaningful and become engaged in the search for it. My aim is to establish a perspective in which

the students and I simultaneously see something meaningful and become engaged in the search for it. It happens easily when the content to be learned responds to a certain interest students already have, but certainly I am not restricted by their existing interests, because my goal is that my students enlarge their cultural interests and build concepts about the essentials of science. To paraphrase Piaget:

It's a question of establishing situations that, not spontaneous in themselves, induce spontaneous thoughts, sparking off students' interest in a knowledge field.

Before students engage in the astronomy activity I planned, they can explore as they use their senses to learn about the phenomenon with minimal guidance.

(2) **Exploiting students' pre-knowledge**. I am alert to the ideas my students have about the concept I plan to develop. To quote Ausubel: "The knowledge students already possess is the most important thing in learning." The presence of misconceptions and conceptual gaps must become explicit. I engage students in active behavior and try to turn their mistakes, misconceptions, and dissatisfactions into problems they will want to work on. When doubts become explicit, new questions arise. Students are guided in looking for significant questions and opportunities of research.

Investigating students' pre-knowledge makes explicit the misconceptions, epistemological obstacles, and gaps. The activity "A round world" is an example focused on the shape of the Earth and gravity. Everybody knows that the Earth is a sphere, but the majority of people don't use the concept of Earth-sphere. By drawing a picture of "the Earth, clouds, and rain," students are guided to explore their implicit conceptions about a flat or spherical Earth and about their idea of up and down. Often students are not able to connect the spherical form with its geometrical and physical properties: it appears evident when a round Earth is sketched in space, clouds are on the upper part, the rain falls off the Earth to an absolute down in space, independent of the gravitational force (Fig. 6.5). Perhaps we haven't come very far from the times of Plinius, *Historia Naturalis*:

Usually we talk about the Earth's sphere, but the struggle between science and popular opinion is great. They say that men stand all over the world, and their feet are opposed to each other. If matters stand thus, why don't the people at the antipode fall from the Earth? And they probably wonder why we don't fall in the same way! Another question arises: why the Earth itself is hanging in the sky and doesn't fall with us?

By showing how the human mind has evolved throughout history, I give students the elements of a critical look.

(3) **Giving preference to questions over answers**. It is best to develop an inquiry attitude using significant – sometimes "upsetting" – questions. Common-sense rules and scientific knowledge don't provide all the answers. In this uneasiness, new knowledge links with old knowledge in a more conscious way.

An example concerning orientation: "Knowing the direction south–north, where is Norway? Imagine that the Earth is glass and you can look through it. Which way would you look, in a straight line, to see that place? Which way would you point?"

In the northern hemisphere, popular opinion associates south with down and north with up, so most students point off into space. In that direction there is the polar star. Though Norway is northward, its location is downward with respect to the local horizon. The Earth is a sphere and "everybody feels on top of that." The whole Earth is under our feet.

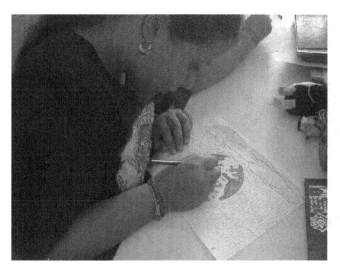

Fig. 6.5. Investigating students' conceptions about the shape of the Earth and gravity.

(4) **Caring about linguistic aspects**. The previous example shows the importance of connecting "common" with "scientific" language because concepts are expressed by words and often one word has several different meanings. For instance, with the same word "north" is meant:
   a.   the direction of shadows at midday is the north direction on the local horizon;
   b.   the North Pole of the Earth is the geographical North Pole;
   c.   the direction of the polar star is the celestial North Pole;
   d.   the magnetic North Pole.
   a, b and c are on the same meridian plane, but at different heights on the horizon:
   a.   in the plane of the local horizon;
   b.   below the local horizon;
   c.   above the local horizon.
   Magnetic North Pole isn't on the same meridian plane.

(5) **Practicing mathematical skills**. Calculating, ordering, classifying, comparing, measuring, recording, graphing, reasoning, inquiring are essential skills for studying astronomy; also, astronomical topics can be integrated into mathematical lessons.

(6) **Giving time for understanding from observations**. Students should gain experiences before working on more theoretical explanations. They spend a long time in observations and hands-on activities (Fig. 6.6) before developing models and theoretical explanations. I want students to discover the ideas of astronomy for themselves, not just to read about them passively. Active participation renews emotions, curiosity, and effective learning. The act of discovery, of experiencing insight or understanding, produces a thrill of victory, and its memory lasts forever. In search of reference systems, we start observations from our local horizon. Students first discover how the sun changes in its path each day and throughout the year and then they create a concrete model to understand the seasons. They observe and record where the sun sets (or rises) over one year, then the skyline becomes a horizon calendar and the sky a scientific laboratory, always available, free and fully equipped.

Fig. 6.6. Hands-on activities.

(7) **Collecting data in order to discover simple general laws**. This includes data such as the midday height and declination of the sun throughout one year. Once a week, at midday, students "catch a sun ray" and measure the height of the sun with a quadrant. Creating a complete record of data would take one year, but in just six months students can foresee further results: the height of the sun at midday, on the equinoxes, is the complement of latitude; on the winter solstice the complement of latitude −23°; on the summer solstice the complement of latitude +23°.

(8) **Starting nearby and moving further out**. This strategy is used to determine distances in the universe from Earth to moon, sun, planets, stars, and galaxies. Students have to be guided to follow humanity in its discovery of the shape of the Earth, reference systems, celestial motions, distances in the universe, measure of time. Gaining knowledge about distances in space was hard and long work, and a step-by-step process. Students work, on those steps, using historical methods and data gained by observations and can see that astronomers begin from simple principles to make deductions about the scale of the universe.

(9) **Building and using models**. Models can add a lot to the understanding of astronomical concepts in order to develop geometric and spatial imagination and link celestial phenomena to terrestrial analogies. Many models can be constructed from common materials, hand made and manipulated by students, but other model types I found very successful are "dynamic human models." Effective learning of astronomy at the primary level is strongly related with an active body: when students enjoy a model in which they cooperate with others, the retention of concepts is enhanced. These models need a long time to arrange and enact, and this time helps students to reach the spatial visualization required for many astronomical concepts. Some examples:

- A *human body gnomon*: a human body is used as a gnomon for building analemmatic and tangent sundials.
- *Planets in motion*: five students (visible planets) stand in large concentric circles with the sun in the very centre of them. The circles represent the planetary orbits (not to scale) and are marked with as many stones as it takes Earth months for the planets to revolve around the sun (3 for Mercury, 7 for Venus, 12 for the Earth, 23

Fig. 6.7. Human models: "the week's dance" and the planetary week.

for Mars ...). Twelve points along the circle of horizon are zodiacal constellations. The planets have to walk around the sun with the same rhythm, at regular intervals of time, step by step, stone by stone. All students are invited to take a turn playing the role of the Earth and thus can see the planetary motion from a point of view internal to the model. In this way they discover easily why, how, and when the real planets appear to wander against the stars for an observer on the Earth.

• *Walking with solar steps*: this dynamic human model helps students to understand the rhythm of the sun motion throughout the year, how and why days stay long (and nights short) for several weeks near the summer solstice, and how and why they shorten fast before the winter solstice. We work with the monumental sundial in St. Peter's Square in Rome: the gnomon is a giant Egyptian obelisk, and the local meridian is lined with zodiacal signs. The midday shadow of the obelisk passes always in one month from one sign to the next one, but it covers very different distances. The question is: "how can the sun cover different intervals in the same time?" In order to answer, students "walk with solar steps": always 30, to correspond to the days in a month, but the steps have different length: in summer, snail's paces, in winter, very long paces.

• *The planetary week and the "week's dance"* (Fig. 6.7): active physical participation, rhythm, and movement help students to understand the reason of the sequence of the weekly days. This model enacts the explanation of the origin of the planetary week made by Abraham ba-Hiyya (1065–1136), from Barcelona, called Savasorda, found in *Sefer ha-Ibbur* (Book of Intercalation), written in 1122–1123, which is the first Hebrew work devoted exclusively to the study of the calendar. It started from the division of the day in 24 equal parts, hours. Each hour is dominated by a planet (luminary: Saturn, Jupiter, Mars, the sun, Venus, Mercury, moon). Each

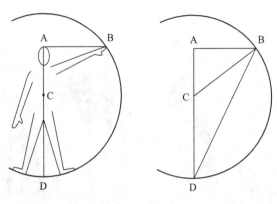

Fig. 6.8. Discovering proportions and laws in the human body.

day takes its name from the planet that dominates its first hour (regent). We work in the outdoors. Seven students, corresponding to the seven luminaries, are placed along a circumference, according to the Ptolemaic system, and they count the 24 hours of a day walking along (and inside) the circle. After 24 steps, everybody reaches the reigning planet the next day. It's a dance; movements can be slow or fast, but synchronized, and a star with seven points appears. It can be marked with a colored string.

(10) **Making students aware of the path they are following**. In the first lessons, students are involved in a guided tour through an unknown field, the map planned by the teacher; gradually they are going to discover and know several topics in that field and build their own conceptual maps. At the end of every year, an interactive exhibition documents and focuses on the main significant concepts students have worked on. Students train other students in the same activities and learning situations they have experienced. It is a summary as well as a test of the effectiveness of their understanding and learning.

6.4   An interdisciplinary approach

Among the most distinctive aspects of my teaching astronomy there is an interdisciplinary approach. A deep emotion and desire of knowledge emerges from the sky: physics, mathematics, geometry, science, art, religion, literature, music, without separation into different disciplines. This integration has to be emphasized at the elementary school level.

Astronomy fits into a wide range of human culture:

- **Anthropometry**. This subject involves measuring the human body in order to discover proportions and laws (Fig. 6.8). Inside the astronomy course I developed a special course, "The laws of the universe inside us," which starts with the measurements of the human body in order to guide 11- to 14-year-old students – teenagers becoming aware of their physical changes – to use mathematics and discover general rules in nature, to achieve a scientific method, and to understand the meaning of laws in physics. As the work goes on, they find out that regularities can be found both in the other aspects of nature and in art – one proportion is the Golden Ratio – and they are ready to understand the words of Galileo: "Nature is an open book, readable for those who know the language of mathematics."

Fig. 6.9. Secchi's observatory in Collegio Romano.

- **Anthropology**. We investigate the astronomical roots of our culture, the astronomical phenomena at the basis of our calendar, customs and feasts. At solstices and equinoxes, during workshops and observations, students set tables with seasonal decorations and foods, and tell the traditions, customs, rites, which are still alive in their families and countries.
- **Mythology**. Myths describe the origin of the universe, planets and constellations, and help to organize the space of the sky. Myths provide the first sky-maps. The names of the planets are taken from ancient Roman and Greek mythology; in parallel with learning about the planets, students can learn the mythological myths and subjects. The same possibility exists about constellations. These mythological subjects have attracted the interest of many artists and they can restore the leading role of astronomy in education.
- **The history of astronomy**. In my teaching, history has an appropriate place. My major motivation in teaching segments of history of astronomy is to make students aware of the struggles innovators go through in order to produce changes as well as to demonstrate that astronomy has had contacts with a lot of other aspects of life for many centuries. We trace back the origins of the early astronomical observatories. In Rome they are connected with the history of the Jesuits who were in charge of astronomical research until the last century. We hold workshops and lessons in Collegio Romano (Fig. 6.9) and in places visited by Galileo, to train students into the history of astronomy.
- **The sky and the cities**. The social and ritual foundation of a city has always been related to the sky: we carry out research concerning how ancient architects organized place and time by observing astronomical phenomena. Archaeology enters our lessons

of astronomy. Ancient monuments have revealed a much greater astronomical interest and sophistication among ancient civilizations than standard histories give them credit for. Monuments, historical buildings, squares become our laboratories. In Rome, the Pantheon, the Domus Aurea, and the Villa Adriana are examples of suggestive places in which the history of a city, the project of an emperor, astrological symbols, and astronomical data fuse in geometric architectural structures. Some monuments work as sundials and calendars.

- **Monuments like instruments**. We use historical monuments like instruments to register astronomical phenomena. Several monuments work like a sundial and a calendar. We work inside them and reproduce similar instruments at school. Examples: The sun in the temple: the Pantheon in Rome as a solar observatory (Figs. 6.10, 6.11). The sun in the church: Ste. Maria degli Angeli and its sundial, astronomy in St. Peter's Square.

6.5   To promote effective astronomy education

The main conditions that helped me in the development of the astronomy course are my enthusiasm for astronomy, my interest in education research, and my integrated teaching of mathematics, science, and astronomy. But, in order to increase effective teaching of astronomy on a large scale, an effective teaching education is essential. Already in the third century BCE the Greek astronomer Arato da Soli wrote in his *Phenomena* that for looking at the sky you need an inspiring guide close near you, pointing to the stars. This means that, to enjoy learning about astronomy, students need enthusiastic and knowledgeable teachers to guide their explorations, teachers interested in learning along with them. Even teachers have to experiment with inspiring and effective learning situations, to find in the sky a fascination and source of a deep desire of knowledge, to discover that a surprising abundance of observations is possible in the outdoors, to assume an inquiring attitude, to adapt their knowledge to the requirements of their students, and to discover their process of learning. Adults and children learn differently: adults have a more organized background, different motivations, and more awareness of their learning, but they can both enjoy the same outdoor observations and experiences. Light is the main agent bringing information from the universe. So natural light plays an important role in the curriculum dealing with observational astronomy. For young students, before they come across a theoretical description of the nature of light, it is essential to start with phenomena they can observe in everyday life: light and shadow. For adults, as well as for pupils, light and shadow can be a surprising topic (Fig. 6.12; Fig. 6.13)!

Teaching is effective – and *knowledge becomes culture* – when it is useful for understanding the reality and for acting on that. It happens when teaching comes across the individual process of learning by significant aims, methods and subjects. It happens with astronomy, when it helps students:

- to develop space/time concepts, essential for understanding the world;
- to adapt their education to the requirements of everyday life;
- to find that what they learned at school is to be found again and again beyond the school walls. Astronomy can make a great contribution to education: let's use this opportunity!

Fig. 6.10. Rome, the Pantheon in summer.

Fig. 6.11. Rome, the Pantheon in winter.

Fig. 6.12. Children's activities in light and shadow.

Fig. 6.13. Adults' activities in light and shadow: how many people in a human shadow?

Further reading

Davidson, N. 1985, *Astronomy and Imagination*, London and New York: Routledge and Kegan Paul.

Fucili, L. and Lanciano, N. 2000, *Il cielo a scuola Naturalmente anno*, **13**(1), 56–8.

Fucili, L. 2001, "Astronomical terraces: space labs between the Earth and the sky." ESA Teach Space.

Fucili, L. 2002, "Building 'meaningful paths' for teaching science." Proceedings Woudschotenconferentie, *Learning Lines in Teaching Physics*, Noordwijk: ESA.

Lanciano, N. 2002, "Strumenti per i giardini del cielo," *Edizioni junior*, Bergamo.

Comments

Robert Stencel: To all speakers: how are you dealing with the problem of light pollution in your curriculum?

John Percy: The International Dark-Sky Association and the IAU's Commission 50 are actively developing education and outreach resources to help us all. Reference: www.darksky.org; www.ctio.noao.edu/light_pollution/iau50/

7

The *Astronomy Education Review*: report on a new journal

Sidney C. Wolff and Andrew Fraknoi

(Wolff) National Optical Astronomy Observatory, Tucson, AZ, USA
(Fraknoi) Foothill College, Dept. of Astronomy, Los Altos, CA, USA

Abstract: The *Astronomy Education Review*, "a lively compendium of research, news, resources, and opinion" about astronomy and space science education, has now been in operation since 2002. This paper discusses why such a journal/magazine was needed and describes the key decisions that were made about how to implement it: electronic publication to make the journal widely accessible while keeping operations costs low; achieving quality comparable to a paper journal through rigorous refereeing, copy-editing, and consistent style; fixed text after publication; and long-term availability. The topics covered in the issues published to date are summarized, and we assess the impact of the journal. The journal itself can be found at http://aer.noao.edu.

7.1   Introduction

The *Astronomy Education Review (AER)*, a new electronic journal, was established to serve the growing community of researchers and educators who are active contributors to astronomy and space science education.

For decades, it has been a cause for concern that too few US undergraduates are being trained in science, engineering, and mathematics (NCEE, 1983). In addition, our increasingly technological society requires citizenry that is at least somewhat literate in science. As a consequence, National Aeronautics and Space Administration (NASA) and National Science Foundation (NSF) have committed very significant resources toward improving both access to, and the quality of, science education at all levels. Proposals to the NSF are now judged not only on the intellectual merit of the proposed research but also on its broader impacts, including how well the work will advance "discovery and understanding while promoting teaching, training and learning" (NSF, 2004). And one of the three overarching missions that underpin NASA's current strategic plan is "to inspire the next generation of explorers. . . . as only NASA can" (O'Keefe, 2002) and, specifically, to motivate students to pursue careers in science, technology, engineering, and mathematics.

Astronomers have long argued that, because of its broad public appeal, astronomy has a unique role to play in education and public outreach, and that it is an extremely effective subject area for attracting young students to the study and enjoyment of science, whether or not they ultimately pursue it as a career. A long-standing commitment among astronomers to engaging both students and the public in the excitement of astronomical discovery, combined with the substantial investments in education by NASA and NSF in the USA (and similar agencies worldwide), has produced a growing cadre of professionals trained in both astronomy and education and whose primary professional activity is education in astronomy and space science. Many other astronomers are actively involved at least part-time in public outreach and in education at all levels from K–12 through graduate school.

It has been recognized for some time that the astronomy education community needs enhanced support from the kind of infrastructure that is typically available to any healthy profession (Fraknoi, 1996). For example, the most recent Astronomy and Astrophysics Survey Committee (AASC) survey (2001) of priorities for the first decade of the new millennium recommended that:

The engagement of astronomers in outreach to the K–12 community should be expanded and improved by ensuring:

(1) appropriate incentives for their involvement;
(2) training and coordination for effective and high-leverage impact;
(3) careful scrutiny of major initiatives and widespread dissemination of information regarding their successes and failures; and
(4) recognition of the value of this work by the scientific community.

A peer-reviewed journal can support many of these goals. A journal is the mechanism by which most professions communicate information, validate quality, and build a growing database of knowledge. Acceptance of manuscripts for publication in peer-reviewed journals is a key method of recognizing achievement among a profession's practitioners.

Most sciences, including physics, chemistry, and geology, already have journals and/or magazines devoted to educational issues. Astronomy education, however, until recently, seemed too small and disjoint a field to make any kind of journal financially practical. Nevertheless, since the need for such a journal was becoming clearer (Hemenway, Roettger, and Percy, 1996), discussions of possible approaches took place at a number of meetings and meeting sessions devoted to astronomy education in the 1990s.

Eventually, an ad-hoc committee to look into founding such a journal was formed with representatives of the American Astronomical Society (AAS), the Astronomical Society of the Pacific (ASP), and the Astronomy Education Committee of the American Association of Physics Teachers (AAPT). The committee was chaired first by Juan Burciaga and then by one of the present authors (Fraknoi). An informal web-based survey taken by this committee showed strong interest in a journal, but produced few practical ideas on how to bring one about. There was much debate on the committee about ways to start such a journal, but the discussion foundered on the issue of how to support the publication on an ongoing basis. Still, the creation of a peer-reviewed journal devoted to astronomy education was endorsed by the Astronomy Education Board of the AAS in its 1999 Long Range Plan, which called for investigating "the establishment of a joint, refereed, electronic journal on astronomy education (together with [other societies])."

7.2   The launch of *AER*

Both the staff time and the resources to launch an astronomy education journal became available when one of the present authors (Wolff) stepped down as Director of the National Optical Astronomy Observatory. NOAO is funded primarily by the NSF and therefore has support of education as one of its missions. An education journal is the kind of high leverage activity that is appropriate for a national observatory.

The key decision that made *AER* possible was the commitment to publish electronically only. NOAO already has in place the infrastructure needed to support electronic publication. The observatory maintains a robust website and has staff who are expert in both web design and html programming. NOAO already takes in and tracks observing proposals, and much of the software developed for that purpose can be used for receiving, tracking, and formatting manuscripts with only incremental modifications. The start-up costs were therefore quite low and involved primarily the time of the two editors along with 0.5 FTE paid staff, primarily for copy-editing and formatting manuscripts. Publicity for the journal (including notices to magazines, bulletin boards, exploders, potential authors, etc.) can all be done electronically, and the only cost is the time involved in assembling distribution lists. NOAO supported most of the programming required, which to date has been only a few person-weeks of *AER*-specific work. Another enabling factor was that the programming for *AER* could be flexibly scheduled around time-critical observatory priorities such as handling observing proposals.

One prerequisite for the continued operation of the journal is that the costs remain as low as possible. Over the next two years, we will be further automating many of the processes used to track manuscripts and format them for publication. The goal is to operate the journal with part-time editors, the active assistance of the editorial board in finding and corresponding with referees, and a half-time copy-editor. We believe that copy-editing is essential in order to maintain the quality of the journal.

7.3   Electronic publication

Electronic publication was chosen for *AER*, both because it is the lowest cost option and because the community of expected readers is quite web-savvy. Paper journals generally derive their income from subscriptions, page charges, and/or advertising. Because we are committed to ensuring that the journal can be accessed by everyone with an interest in astronomy education, we do not plan to charge either readers or authors. Advertising alone could not generate a sufficient income stream to support the costs of paper publication.

Electronic publication offers a number of other advantages relative to paper publication. First, the speed of publication can be higher. We currently post papers to the website as soon as they are refereed, revised, and copy-edited, rather than waiting until we have enough articles to complete an entire issue. An electronic journal can be easily searched for specific information. Image formats are flexible, and diagrams and photographs can be published in color at no increment in cost. It is easy to download only those articles of interest. And perhaps most important of all, electronic publishing enables worldwide access (including readers in countries where budgets do not permit subscriptions to journals).

In formulating the design of *AER*, we wanted to make it clear that we were publishing a journal, not merely developing a website. As much as possible, we have tried to give *AER* the look and feel of a journal. We adopted a very conservative design, which has the additional advantage that it downloads quickly on all browsers. All of the papers can be printed in PDF format. At the conclusion of each issue, we assign page numbers to the papers so that they can be referred to in bibliographies in the same format as used for paper publications. Contrary to our early expectations, authors have not been reluctant to submit quite long papers for electronic publication.

7.4   Content

In determining what kinds of papers *AER* would publish, we made the judgment that there were unlikely to be enough astronomy education research papers to sustain the interest of readers. Furthermore, the community of astronomy and space science educators needs more than research results – it needs more awareness of ongoing programs, available resources, and new opportunities. Accordingly, we used as our model journals like *Science* and *Nature*

rather than the *Astrophysical Journal*. In addition to publishing research papers, we wanted
to include in more of a magazine format a variety of other types of contributions that would
be useful to practicing educators. As a consequence, *AER* has five sections:

> *Research and Applications*: refereed papers and review articles on research in astronomy
> education, along with ideas about how to apply results of such research in "real life."
> *Innovations*: short reports on innovative techniques, approaches, activities, and materials.
> These reports convey the essence of the innovation and include ways that interested
> readers can learn more.
> *Resources*: annotated lists of useful resources for any level or arena of astronomy edu-
> cation.
> *Opportunities*: short announcements of funding sources, cooperative projects, employ-
> ment (jobs that are 50 per cent or more education), workshops and symposia, etc. In
> most cases, such announcements are one or two paragraphs, with a weblink for further
> information.
> *News, Reviews, and Commentary*: letters to the editor, editorials, resource reviews, opin-
> ion pieces, and interactive discussions.

Papers published to date have covered topics related to all levels of formal education from
primary grades through graduate school. We are also interested in receiving articles on work
in informal education and public outreach. Topics covered in the articles published to date
include techniques of assessment and how the various techniques affect student learning;
the effectiveness of collaborative learning groups; student misconceptions; scientists' mis-
impressions about the national education standards; the evaluation of Internet and distance
learning courses; the background and application of the astronomy diagnostic test; how to
make astronomy accessible to students with learning disabilities; a review of the literature
in astronomy education research; a report on workshops that developed a list of goals for
introductory astronomy for non-science majors; resources for using poetry and science fiction
in astronomy courses; a number of opinion pieces; and much more.

7.5   Impact and quality

In order to be judged a success, a professional journal must be read, publish high-quality
papers that advance the field, and enhance the professional recognition of the contributing
authors.

AER was formally announced at the meeting of the American Astronomical Society in
January 2002. We publish one volume each year, with two issues per volume. The number of
accesses to *AER* has shown a steady overall increase with time, apart from drops during the
summer months. In each of the two most recent complete months (October and November
2003), *AER* received over 210,000 hits from more than 7,700 different IP addresses, with an
average of 16,000 article downloads each month.

AER has recently asked readers to register so that we can obtain some demographic information about them. Registrants will receive a free newsletter when each issue is complete,
summarizing the articles contained in that issue. Currently, 25 per cent of the registrants are
from outside the USA; 26 countries in North and South America, Europe, Asia, and the Middle
East are represented. We know from anecdotal evidence and conversations that the papers are
being read not only by astronomy education specialists but also by researchers who teach.

Papers in *AER* have been discussed in journal clubs that previously focused exclusively on astrophysical research papers.

Two factors determine the quality of the published papers: the quality of the initial submission and the quality of the refereeing. It has been very helpful to the successful launch of *AER* that several of the leading researchers in astronomy education decided to publish papers in the first issues of the journal, rather than waiting to see how successful *AER* would be. In other words, they took a chance on us. A handful of papers were solicited by the editors, primarily after seeing poster and oral presentations at meetings. The vast majority of papers have been unsolicited.

All of the research papers in *AER* are sent to external referees. In addition, we have chosen to have all of the papers dealing with innovations and resources, along with most of the other contributions (except short news announcements), refereed by external reviewers, and their comments have led to material improvements in many of the papers. Referees remain anonymous unless they specifically request otherwise. About 25 per cent of the papers submitted to *AER* are rejected either before (by the editors) or after refereeing.

One of the primary questions about electronic publication is whether it will count toward professional advancement. Astronomy does have one advantage over other fields in that we are used to having all of the astrophysics literature available electronically. Since electronic access is taken for granted – and indeed for many of us has become the preferred mode of searching the literature – electronic publication is not a barrier for our readers.

A search of the literature available via the web indicates that most universities and other organizations have not yet established policies for evaluating electronic publications. Discussions of this topic (e.g., Neely, 1999, and references therein) generally conclude that the primary consideration, as for paper publication, should be quality. *AER* is committed to maintaining the same level of quality control as a paper journal. The factors that determine the quality of published scholarly work are refereeing, sustainability, endorsement or support by professional societies, the impact factor, international reach, and coverage by abstracting and indexing services.

How does *AER* measure up when it comes to these factors? We have already described *AER*'s commitment to thorough refereeing. The papers themselves will be long-lived. After publication, the text is frozen. The journal itself is stored at an institution (NOAO) whose website is rarely down and which has backup procedures in place. Copies of completed issues are stored off-site for further protection.

AER was launched with the endorsement of the two largest professional astronomy societies in the USA – the American Astronomical Society and the Astronomical Society of the Pacific. In its most recent Long Range Plan (November 2003), the Astronomy Education Board of the AAS again confirmed its commitment to the importance of *AER* and formulated increased publication of astronomy education papers in the journal as one of its key goals for the coming years.

To add to the journal's credibility and to benefit from the advice of those who have been thinking seriously about the future of the field of astronomy and space science education, we established an Editorial Board and a Council of Advisors for the journal. The 36 members of these two groups include a good cross-section of the leaders in astronomy and space science education. (The names of the Board members are listed in a section of the journal site.)

Both NASA through a specific grant and NSF through its funding of NOAO have contributed financial support to *AER*. The commitment of the two leading funding agencies for research in astronomy and space science is a clear indicator of the importance attached to enhancing the professionalism and recognition of specialists in astronomy and space science education.

The impact of *AER* on classroom practices and on the development of educational programs and materials is more difficult to measure. We do know that the journal articles are being downloaded in substantial numbers, and papers published in *AER* are now being referenced in the literature and so are contributing to the growing body of knowledge about effective practices in astronomy education. Talks at education sessions at astronomy meetings frequently include a discussion of the journal or one or more of its review articles. *AER* is also reaching an international audience. A large number of libraries and astronomy organizations, both inside and outside the USA, provide links to *AER*. We will also explore what is required to be covered by various indexing services.

7.6   The future of *AER*

The long-term success of *AER* depends on its contributors and its audience. All those who are committed to improving the quality of astronomy and space science education can help ensure the health of the journal by:

(1) reading *AER* and discussing relevant papers with colleagues;
(2) submitting papers, articles, and news items to *AER*; we are particularly interested in expanding the content to include more papers relating to K–12 education and more international papers;
(3) volunteering to referee papers;
(4) citing *AER* papers in other published manuscripts;
(5) using the papers in classrooms, graduate student training, journal clubs, etc.;
(6) registering with the journal to give us some demographic information about our readers (and also to receive a newsletter twice each year about the most recent contents).

The *Astronomy Education Review* can be found at http://aer.noao.edu; or by typing "Astronomy Education Review" into Google; or by clicking on Education from the NOAO home page (http://www.noao.edu).

We gratefully acknowledge the financial support of NASA's Office of Space Science and the National Optical Astronomy Observatory.

References

Astronomy and Astrophysics Survey Committee (AASC) 2001, *Astronomy and Astrophysics in the New Millennium*, Washington, DC: National Academy Press, **15**, http://books.nap.edu/books/0309070317/html/15.html

Fraknoi, A. 1996, "The state of astronomy education in the US," in J. Percy, ed., *Astronomy Education: Current Developments, Future Coordination*, ASP Conference Series, **89**, San Francisco: Astronomical Society of the Pacific, 9–25.

Hemenway, M. K., Roettger, E., and Percy, J. 1996, "Creating networks and coalitions," in J. Percy, ed., *Astronomy Education: Current Developments, Future Coordination*, ASP Conference Series, **89**, San Francisco: Astronomical Society of the Pacific, 143–7.

NCEE (National Center for Education Evaluation) 1983, *A Nation at Risk* http://www.ncrel.org/sdrs/areas/issues/content/cntareas/science/sc3risk.htm

Neely, T. Y. 1999, http://www.arl.org/diversity/leading/issue10/tneely.html

NSF (National Science Foundation) 2004, http://nsf.gov/pubs/2004/nsf042/3.htm#IIIA2

O'Keefe, S. 2002, http://www.nasa.gov/about/highlights/Pioneering.html

Comments

Andrew Constantine: What differences did you notice between US expectations and the Turkish paper that was presented to the *Astronomy Education Review*?

Reference

Gurel, Z. and Acar, H. *Astronomy Education Review*, **2**, 1.

Sidney Wolff: The difference in understanding weightlessness was interesting to note; the level of understanding was far poorer in Turkey.

James White: Can you currently, or will you be able to in the future, capture information about the *AER* users? Attempt to determine the groups using the *AER*?

Sidney Wolff: We are currently tracking only the number of hits to, and downloads from, the *AER* website. Although access to the journal will remain free of any charges, we have asked readers to register as subscribers and to supply some demographic information. In return, subscribers will receive email twice each year summarizing the contents of the latest issue. About 25 per cent of the early subscribers live outside the USA, but we have not yet tabulated the other data that we requested about their training and professional activities.

Poster highlights

Astronomy education research in New Zealand was represented by **Bill MacIntrye**, in a poster entitled **3D modeling: demonstrating understanding in astronomy**.

An extensive literature base on alternative notions (misconceptions) held by children and adults about the nature of astronomical events has been collected almost entirely from oral and/or written responses of the participants and not through the use of three-dimensional (3D) models. Parker and Heywood (1998) and Barab *et al.* (2000) found that the use of practical modeling was an effective strategy to develop astronomical understanding. Therefore, if the use of practical modeling is an effective strategy to develop astronomical understanding, then

Bill MacIntyre (New Zealand) demonstrating modeling of seasons at a teachers' workshop in Sydney, July 26, 2003. Photo by Rob Hollow.

one would expect the use of 3D models to be an effective pedagogical approach to eliciting astronomical understanding.

Whether the purpose is to demonstrate understanding or to elicit prior understanding of basic astronomical events, 3D modeling is an effective approach. Students or teachers can provide explanations of astronomical events by physically modeling the evidence that supports their explanation statements. Students demonstrating an explanation for the cause of seasons would use two 3D models of planet Earth (one on a 23.5° tilted axis and the other at a 90° angle) to demonstrate the evidence. Possible evidence that the student might demonstrate with the two different 3D models are: sun high in the sky during summer and low in the sky during winter; Antarctica in full sunlight during the summer and in total darkness during winter; long daylight hours in summer and short daylight hours in winter; etc. Students will see for themselves that only the 23.5° tilted axis model will demonstrate all of the seasonal evidence. The physical modeling of the evidence allows the viewer to discern whether the modeler has an appropriate science understanding for the cause of seasons or is relying on a memorized statement of fact.

References

Barab, S. A., Hay, K. E., Barnett, M., and Keating, K. 2000, *Journal of Research in Science Teaching*, **37**(7), 719–56.
Parker, J. and Heywood, D. 1998, *International Journal of Science Education*, **20**(5), 503–20.

Strategies employed to use astronomy as an introduction to the skills of critical thinking are explored here in **Critical and scientific thinking in astronomy courses** by **Harry Shipman**.

The paper summarizes the results of several studies, many conducted with science education collaborators (Nancy W. Brickhouse, Zoubeida Dagher, and Will Letts). They found that students can and do learn to appreciate evidence, and many learn how to cite evidence to support scientific claims (*Journal of Research in Science Teaching*, 2000; *The Physics Teacher*, 2000).

While students can learn the relationship between evidence and core scientific theories like the Big Bang Theory, their understanding of this relationship depends on the theory, in contrast to previous assertions in the science education literature (*Science and Education*, 2002). Students show particular difficulties in understanding the evidence in support of stellar evolution. Student understanding of the nature of scientific theories is more problematic. However, teaching sequences constructed as a result of this research show some promise (*International Journal of Science Education*; *Astronomy Education Review*).

This paper contains specific examples of teaching strategies used to help students learn critical thinking and scientific habits of mind.

References

Brickhouse, Nancy W., Dagher, Z. R., Letts, W. J., and Shipman, H. L. 2000, "Diversity of students' views about evidence, theory, and the interface between science and religion in an astronomy course," *Journal of Research in Science Teaching* **37**(4), 340–62.
Brickhouse, Nancy W., Dagher, Z. R., Shipman, H. L., and Letts, W. J. 2002, "Why things fall: evidence and warrants for belief in a college astronomy course," *Science and Education*, **11**(6), 573–88.
Dagher, Z. *et al.* 2004, "What is a theory? What is a law?" *International Journal of Science Education*, **26**(6), 735–55.
Shipman, H. L. 2000, "Teaching astronomy through the news media," *The Physics Teacher* **38**, 541–2.

An example of interactive teaching is next compared to astronomy courses that rely exclusively on the Internet for content dispersion in **An online astronomy course vs. an interactive classroom** by **Timothy F. Slater** *et al.*

Slater and colleagues designed a hypermedia learning experience (*Astronomy Online*) for introductory astronomy that matches Internet technology with how people learn. In one approach, the course was delivered via the Internet to middle- and secondary-schoolteachers spread across the globe. In another, the course was delivered to undergraduate non-science majors, where the only class meetings were orientation sessions and on-campus exams. Slater *et al.* compared these with conventional on-campus courses that use interactive teaching techniques by studying common examination questions, the Astronomy Diagnostic Test, an attitude survey, and student interviews. The results obtained from the qualitative instruments and the interviews suggest two overarching ideas. First, students are indeed learning successfully in this online learner-centered environment as evidenced by both the content conceptual test and the Astronomy Diagnostics Test; however, they are not showing gains as large as those in the learner-centered on-campus course. Second, student attitudes toward astronomy and online learning are mildly positive, and do not seem to change much over the semester. They were considered both to be positive conclusions. *Astronomy Online* is available through W. H. Freeman & Company, United States, ISBN 0-7167-9669-4.

Further focus on information access on the Internet is provided by **Guenther Eichhorn** *et al.* in a poster entitled **Access to the astronomical literature through the NASA ADS**.

The NASA Astrophysics Data System (ADS) is used by astronomers worldwide. The searchable database contains over 3,000,000 records. In addition the ADS has scanned 2,100,000 pages from more than 300,000 articles, dating back as far as 1829. There are currently more than 10,000 regular users (defined as more than ten queries/month). ADS users issue 1,000,000 queries, receive 30,000,000 records, and 1,200,000 scanned article pages per month. The ADS is accessed from about 100 countries with a wide range of the number of queries per country. Approximately one third of the use is from the USA, one third from Europe, and one third from the rest of the world. In order to improve access from different parts of the world, we maintain 11 mirror sites in Argentina, Brazil, Chile, China, England, France, Germany, India, Japan, South Korea, and Russia. Automatic procedures keep these mirror sites up-to-date over the network. Both the search system and the scanned articles in the ADS can be accessed through email. Email can be used by users who are on slow or unreliable Internet connections. It allows access to the ADS for users who do not have a good enough connection for a web browser. The ADS is funded by NASA Grant NCC5-189.

The ADS is free and available to anybody worldwide at http://ads.harvard.edu. This provides full text search capability. We also have another new free service, myADS, which provides a personalized service.

Editors' Note: The proceedings of the 1988 Williamstown Conference on astronomy education is now online at ADS. We look forward to having more astronomy education material available in this form.

Sergei Gulyaev, provides a conceptual analysis entitled **Visualization of knowledge: making a map of astronomy.**

Application of a concept of educational "science maps" (Gulyaev and Stonyer, 2002) to astronomy education is developed. Science maps of different scales are illustrated with initial examples exploring the application of this methodology in astronomy and astrophysics.

"Mapping" science draws on the concept of mapping methodology (Novak, 1990), modifying it to provide and illustrate the structure and order in a domain of science knowledge. General systems theory (GST) can be used as a theoretical tool for representation of the levels and hierarchies associated with science-based knowledge (Gulyaev and Stonyer, 2002; Gulyaev, 2003). GST is designed for specifying systems and defining their interrelationships, and also for pointing to gaps in knowledge. What is really important for us, as science and astronomy educators, is that GST provides a framework for conceptualizing the subject. It also provides a framework for teaching astronomy based on the continuity of conceptual understanding, rather than the logical yet reductionist structure of astronomy apparent in many texts or curricula.

GST introduces a hierarchy of levels of complexity for basic branches of science. Developing this approach, we use the analogy with geographical maps, and distinguish the three principal "scales" that are necessary for maps of science. In the astronomical context they are:

- the scale of branches and fields of science and astronomy, where interconnections between various disciplines are shown;
- the scale of theories and hypotheses, encompassing a significant segment of astronomy and astrophysics;

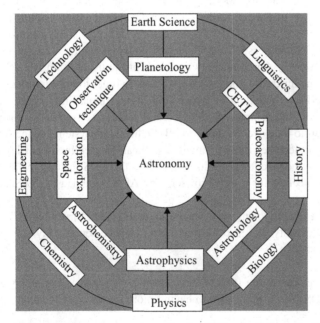

Interconnections of astronomy with other fields of science and technology. Corresponding branches of astronomy arising from these interconnections are specified with linking words.

- the scale of structures and hierarchies, encompassing the geography and anatomy of the material systems and objects essential for a given discipline, as well as principal methods, notions, and concepts it uses.

The accompanying figure illustrates a "scale A" (branches and fields of science) map. Astronomy can be considered both as an independent science with its own tasks, concepts, and methods, and as a derivative science that utilizes information and concepts from physics, chemistry, earth science, and others. Possible interconnections of astronomy with other fields of science and technology are exemplified in the figure. Corresponding branches of astronomy arising from these interconnections are shown as linking words.

It is incumbent upon academics and researchers to conceptualize and define how maps of their fields might be drawn, so that astronomy teachers might utilize them more fully in developing their practice and their students learning in astronomy.

References

Gulyaev, S. and Stonyer, H. 2002, "Making a map of science: general systems theory as a conceptual framework for tertiary science education," *International Journal of Science Education*, **24**, 753–69.

Gulyaev, S. 2003, "New trends in astronomy education: a 'mapping' strategy in teaching and learning astronomy," *ASP Conference Proceedings*, **289**, 151–6.

Novak, J. 1990, "Concept mapping: a useful tool for science education," *Journal of Research in Science Teaching*, **27**, 937–49.

Two posters contributed by **Muglova P. Stoev**, are summarized as **Developing critical thinking skills by active observations in the process of teaching in public astronomical observatories**.

The didactic effectiveness of new methods of involving individual students in educational research activity at the Stara Zagora public astronomical observatory has been investigated. The difference in the level of theoretical knowledge and observational skills of the students from six problem groups of the astronomy school (groups of students involved in the three-year extra-curricular astronomy program) at the public astronomical observatory in Stara Zagora has been objectively determined by tests in the experimental and control groups.

Original programs and structure of the educational content for each of the problem groups have been worked out. The didactic characteristics of the educational cognitive method of astronomical observation have been determined by epistemological analysis. A system of didactic tests for studying the development of the critical thinking skills of the students during their participation in astronomical research observations has been developed and experimented.

A special system of methods for application of the astronomical research observations for development of the critical thinking skills of the students has been formulated. An experiment determining the level of pretensions (the level of ambition as to what he/she feels capable of achieving in the context of problem solving/observation within specified time constraints) of different team observers has been realized. It reveals the individual's preconceptions at objective setting success or failure and the level of observer stability during the specific observation, resulting within the conditions of limited observation time (the duration of the total phase is 144s).

Different versions of textual and observational tests, questionnaires, interviews and the method of expressive appraisals have been used. Two groups of techniques with differences in the procedures and evaluated parameters of the level of pretensions have been applied. In the first group, the students set their objectives in six series of problems graduated in a ten-level scale of difficulty. The main focus in this group is centered on the strategic approach of the students rather than the process of problem solving itself, as the significance of strategic approach is shown in the evaluation of experimental data in this regard. Thus, the strength, adequacy and stability of his/her level of pretensions are evaluated. Ten specially prepared cognitive observational problems have been used as diagnostic material, with which the investigated students have the possibility to be situated in the conditions of real astronomical observational process. The second group of techniques studies the so-called aim/result discrepancy (i.e. the correlation between students' objectives and the factual result). They have no compulsory gradation on the diagnostic material complexity. The observational problems used for the tests have nearly the same difficulty. The relative strength of pretensions, which is an indicator of the students' behavioristic orientation towards achieving a successful end result or not, has been evaluated.

We applied these original methods of determining the level of student's cognitive independence during their training for observation of the total solar eclipses in 1990, 1999, and 2001. We used the didactic system and students from the experimental group have conducted successful scientific investigations of the total solar eclipse on 11 August 1999. The results have been generalized in 11 scientific papers.

References

Stoev, A. 2002, "Creative abilities development conducting amateur research observations in the astronomy school," in *Proceedings of the 4th General Conference of the Balkan Physical Union*, 22–25 August 2000, Sofia: Veliko Turnovo, 156–60.

Stoev, A. 2002, "Visual, photographic, photometric, polarization and other observations during the 1999 total solar eclipse done by the Peoples Astronomical Observatory at Stara Zagora," in D. N. Mishev and K. J. H. Phillips, eds., *First Results of 1999 Total Solar Eclipse Observations*, Sofia: Coronet/Professor Marin Drinov Academic Publishing House, 137–42.

Part III

Educating students

Introduction

A simple curriculum model is cyclic in nature. It begins with aims and objectives. These lead to a choice of content (knowledge, skills, applications, and attitudes). These choices are instilled in the students through effective teaching and learning methods. The teaching and learning are then evaluated and assessed, leading to feedback that is used to improve every part of the cycle.

In previous chapters, we have addressed many parts of this cycle: aims and objectives (implicitly), curriculum content, teacher education, and education research in general. We have not explicitly discussed how teachers should assess students' learning, but we assume that this is part of effective teacher education.

Classroom teaching uses many tools. Elsewhere in this book, we address some of the more innovative ones. There are others, of course, some old and some new. Probably the oldest is the lecture, which can still be effective if it is interactive in the sense of involving questioning – the Socratic approach. There is the blackboard (or white-board) – still a flexible tool. There are the students' notes and notebooks; students can often internalize material by writing it or drawing it. Textbooks have evolved with the times, and now often come with a whole constellation of ancillaries, including a website. Textbooks can be the main support mechanism for both student and teacher.

There are audio-visuals. Wall charts and posters are useful, since they often stay in place for years, and make a permanent imprint on the students' minds. There are overhead transparencies and, looking back a decade or a few, 35 mm slides, filmstrips, and lantern slides. Films, videos, and DVDs (to give an evolutionary sequence) are useful. Nowadays, there is Microsoft PowerPoint and other such presentation software which, depending on how it is used, can combine the best or worst of overheads, slides, and videos. Today's textbook publishers may now even supply their books' illustrations in PowerPoint format.

There are demonstrations or experiments, which may be done by the teacher or by the students. Some of these fall under the category of "hands-on activities" – very effective as long as they are minds-on as well. There are specimens, collections, and museums; meteorites, for instance, always impress students, especially if they can hold them in their hands, both to sense their density and to experience the thrill of holding a space rock in their own hands. Involving students in this way may incorporate a field trip to a museum or science center, where there is specialized equipment such as a planetarium or observatory. NASA's Johnson Space Flight Center even lends out a kit of lunar thin sections, actual bits of the moon for students to examine with geologic microscopes.

There are also investigations and projects, such as for science fairs. These are useful and fun, as long as they are consistent with the fact that astronomy is an observational science, not an experimental one. Students can learn effectively through socialization, which can be done through group projects, discussions, and even through plays, games, and songs.

Television and radio, and newspapers and magazines, may contain astronomical news or other up-to-date content, or may relate astronomy to the real world. Astronomy may also be brought into the classroom by visiting astronomers; this approach can be effective if the astronomer and the teacher are well prepared for the visit, such as through programs like the Astronomical Society of the Pacific's *Project ASTRO*. The American Astronomical Society's Shapley Lecture (www.aas.org/shapley, part of the www.aas.org/education website) program brings visiting astronomers to colleges and universities that lack astronomy programs. They not only interact with faculty and students, but also often consult on improving local infrastructure for astronomy, which may lead to the hiring of new astronomy faculty. The Internet, of course, provides access to almost anything, for better or worse.

8

Textbooks for K–12 astronomy

Jay M. Pasachoff

Williams College, Hopkins Observatory, 33 Lab Campus Drive,
Williamstown, MA 01267, USA

Abstract: I report on American textbooks for kindergarten through high-school grades. Middle school, up through approximate age 15, is the last time American students are required to take science, and I provide statistics on the narrowing of the funnel containing those taking physics. I describe some recent curriculum and standards projects, and discuss the recent "less is more" trend. I conclude with comments on whether textbooks are necessary and useful, and discuss possible content and style of an ideal textbook. Astronomy is orphaned in many American schools, though it can find its way into classes through earth science or physical science courses or textbooks. We can hope that the students will wind up with better astronomical knowledge than Harry Potter, who "completed the constellation Orion on his chart" in June from his own telescopic observations during his practical astronomy exam in *Harry Potter and the Order of the Phoenix* (Rowling, 2003; Pasachoff, 2003a), an observation that can never have been made.

8.1   Standardized testing and the scientific funnel

A national trend in the United States since 2002 has been an increase in required standardized testing, often as part of the No Child Left Behind Act (NCLBA). The unforeseen consequences of NCLBA seem to be an increasing rate of failures and the abandoning of topics of secondary importance – like astronomy – next to reading, writing, and arithmetic.

In New York State, for example, the percentages of students passing the 2003 mathematics exam required for high-school graduation was so far below 50 per cent that the test had to be withdrawn, though similar problems with the physics exam did not lead to a similar temporary solution (Winerip, 2003). One can hope that better training of teachers and clearer expectations can improve this situation in the long run.

The National Center for Educational Statistics has provided graphs that show that, most recently for 1999, the percentage of students who have taken science courses through the age of 17 is 88 per cent for general science (the middle-school course), 93 per cent for biology (usually the first course in high school), 57 per cent for chemistry, and only 17 per cent for physics. Since high-school astronomy is most closely related to physics, the magnitude of the problem is obvious.

The No Child Left Behind Act became politically controversial during 2004, an election year in the United States. It is an example of an "unfunded mandate," where the Federal government requires something to be done but does not provide the necessary money to accomplish the task. Further, its wording requires all categories of students to make improvements each year, so if one out of a dozen categories stays even in testing, the school is labeled "failing," something that is not only inaccurate but also displeases its students and

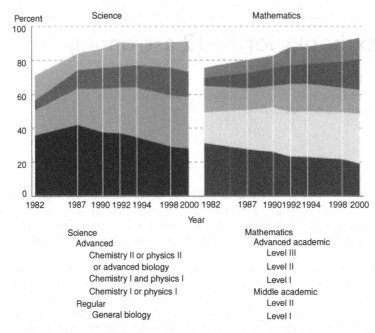

Fig. 8.1. US high-school graduates according to the highest level of advanced mathematics and science courses taken (http://nces.ed.gov).

their parents, not to mention the teachers and school administration. *Editors' Note*: As of Spring 2004, some modifications to the act were being allowed.

8.2   The student sieve

In the United States, all middle-school students (approximate ages 12–15) take science, often a general science course that mixes up topics or a Life Science/Earth Science/Physical Science three-year series. Many or most students don't take any science courses after that. Further, females and minority students disappear from the science courses at a greater rate than average.

Many theories have been advanced to explain this sieve and to try to counter it. Leon Lederman, the physics Nobelist who has devoted much time in recent years to school education, is trying to advance a physics-first order for high-school science, since physics is at the basis of understanding. He is gaining converts. He has been less successful with his attempt to make a scientist the hero of a popular television series, with the funding support his group can provide not proving sufficiently attractive to Hollywood.

8.3   National standards

Education in the United States is not centralized. Unlike the situation in France, where all students throughout the country should be studying the same things on the same day, each division in the United States – often down to individual-school level – chooses its own curriculum. A few large regions – notably the states of Texas and California – provide a list of approved textbooks, sometimes a very short list, giving these states' requirements high profiles in the views of textbook publishers.

Several projects in recent years have tried to provide curricular guidance. The best known is "Project 2061" of the American Association for the Advancement of Science, run as an independent fiefdom within that organization. (The stated hope is that students will become science literate by the time Halley's Comet returns in the year 2061.) They deny that they have provided a curriculum, but they do provide a series of booklets and other materials giving lists of appropriate materials for various grade levels. My objection to their work has to do with the idea that material they consider too abstract or advanced – such as considerations of galaxies – shouldn't be given to students before the ninth grade. Indeed, since the sieve is well advanced by ninth grade, this decision means that younger students don't get to study some of the most interesting aspects of astronomy.

National science education standards have been advanced by the National Research Council, though these are not binding standards. See http://www.nap.edu/readingroom/books/nses/html/action.html.

The National Science Teachers Association also has advanced science standards. See http://www.nsta.org.

8.4   Aids for teachers and teacher training

A major problem in United States science teaching is that most teachers of science have little or no training in science. Indeed, for reasons of salary or prestige, few of the best graduating university students go into pre-university teaching. The students who do go into teaching may graduate from teacher's colleges with many more courses in pedagogy than in subjects. Of course, there are many highly qualified high-school teachers, with some of those in physics equal in abilities to anybody anywhere. But the trained teachers of science are in a minority.

As a result, science textbooks in the United States have elaborate teachers' materials that come with them. Sometimes there are bound volumes with the whole students' version printed at a reduced size to allow for wide margins with answers to questions, hints for teaching, hints for labs, and other material to aid untrained teachers. Sometimes the students' version may not even be reduced, leading to an oversized book for teachers.

The problem of teacher training has no easy solution, and many people are considering it. In the current economic climate, there are no obvious sources of funds to right the many wrongs. Indeed, most funding for American education is local, often funded by a property tax. In New York State, the highest court ruled that "a sound basic education" must be provided for students and, as I write, people are fighting over what that level is. The court ruled that the "systemic failure" of the New York City schools must be overcome and that the State must provide more funding and remedy the situation by July 2004, a deadline that was not met.

8.5   Hands-on activities

Much education research indicates the importance of having students participate in the learning activity. This concept is related, after all, to the difference between teaching and learning: students don't necessarily learn what we teach. Kits of laboratory materials may also be available from the school publishers. Usually, there will be a hands-on activity or two of varying depth in each chapter of a textbook.

8.6   "Less is more?" Astronomy education research

There are many valid objections to the way that many textbooks are chock full of topics with too many ideas and definitions for students to comprehend. This has led to the oft-stated idea that "less is more:" that concentrating on a handful of ideas at length (or in depth) is better for student learning than an encyclopedic approach. But for the bulk of students, it seems probable that a happy medium exists, with enough topics to provide science literacy while leaving ample time per topic. We must guard against less becoming less.

As Partridge (2003) writes, as his valedictory on retiring as Education Officer of the American Astronomical Society (AAS):

Why do I not describe the emergence of a cadre of professional astronomy educators as an unalloyed good? Here are some concerns.... The danger that the educator may drive out the astronomer. If we become so wrapped up in how students learn that we forget what they should be learning, we risk cutting ourselves off from the roots of our field and from our own backgrounds in astronomy. If, for example, we insist that every student must clearly grasp exactly why the moon has phases, we may miss out on the time and opportunity to help students understand larger issues like the evolution of physical systems, be they stars or the Universe. This is the flip side of an exclusive focus on content that has made so many 'Astro 101' courses ineffective. Balance is needed.

See also Pasachoff (2002, 2003b) for comments on phases and seasons that apply to both K–12 education and college education.

An increasing trend among professional educators and physicists devoting themselves to "physics education research" has led to a spin-off of "astronomy education research." The first journal devoted to the topic, *Astronomy Education Review*, has been set up under National Optical Astronomy Observatory auspices by Sidney Wolff and Andrew Fraknoi (see Chapter 7 of this volume). It is an online, refereed journal, available to all at http://aer.noao.edu.

The philosophical questions of how to tackle education at the crucial time of middle school, grades 6–9 (ages approximately 12–15), have been widely discussed, and are related to a conservative/liberal split in the American population. Books such as Hurd (2000) and articles such as Gagnon (1995) typify the actions and reactions.

8.7   Current textbooks

Science textbooks for middle-school years sell in the millions for each of several publishers, so there is much reason for the publishers to devote energy to the topics. Fewer science textbooks are sold for the elementary-school years, though there are age-appropriate books for each grade level. And high-school physics is a minor player on the course scene. Astronomy is taught in very few high schools as a separate topic. In middle school, it is usually part of the Earth Science curriculum and less often part of Physical Science.

I have participated in several K–12 projects, originally the *Physical Science* and *Earth Science* texts of Scott, Foresman and Co. (Pasachoff, Pasachoff, and Cooney, 1983a,b, 1989a,b). I was able to increase the astronomy content, and to move some of it into *Physical Science* from the other book. These two books, along with *Life Science*, took up grades 7–9. We were subsequently allowed to write astronomy chapters for grades 4–6 of the accompanying elementary-school science texts (Pasachoff *et al.*, 1989c). There is at least some astronomy content in each volume of grades K–9, except for *Life Science*.

Though I have tried to add as much modern content as possible, the content of elementary-school and middle-school texts, on the whole, emphasizes facts or concepts like phases and seasons. Little or none of the content in most texts reflects topics of current interest, though even second graders ask visiting scientists about black holes. Some of this is determined by the overlapping and sometimes contradictory published requirements of the states. The limitation often also stems from the application, and misapplication, of sometimes oversimplified psychological theories of Piaget and others about the acquisition of learning skills. The battle over "tracking" of students of different abilities and of "mainstreaming" of poorly performing students also plays a role in the content of texts. The number of American publishers has been diminishing, as publishers merge or are taken over. Only a handful of middle-school science texts remain.

Prentice Hall tried revamping their middle-school series into a set of a dozen individual-topic books, including *Science Explorer: Astronomy* (Pasachoff, 2000, 2005). Since that series was issued, though, the individual chapters were reshuffled and re-released as general science books, sometimes with special content for individual states. A second edition was published, and increased information about space exploration was incorporated; just what to say about NASA and the space program is an interesting question for a book that will be in students' hands for years to come. *The Textbook Letter* (www.textbookleague.org) provides a conservative view of textbooks in various fields, occasionally including science. See also McClintick (2000) on the textbook selection process.

8.8 The accuracy of and readability of middle-school textbooks

Middle-school texts are usually written by a committee, with the editors – themselves often former middle-school teachers – providing detailed lists of what must be included. These lists are based in large part on the overlapping requirements and standards of the 50 states plus other geographical divisions. Though, some decades ago, authors of school textbooks were provided with royalties, as happens with university texts, the trend has been to phase out royalties and to provide flat fees to writers as "works for hire." This change may have negative consequences on accuracy and currency as fewer active scientists are involved. Further, the conglomeratization of American and, indeed, worldwide business has taken over the textbook market as well, so there are many fewer textbook publishers than before.

Unlike the situation with university-level texts, the authors of middle-school texts do not have final approval of the text. Different editors interact with authors on different levels. Indeed, a complaint often seen is that the authors' names are mere window dressing, with the books written by others. I myself have not seen this phenomenon. But it is apparent that some books are written by specialists in education rather than in the subject matter, with a resulting loss of accuracy.

John Hubisz of the physics department of North Carolina State University has become noted for his analysis of inaccuracies in middle-school texts. Some of his reports are on line at http://www.science-house.org/middleschool. They have gained widespread publicity. While many of the items on his lists of inaccuracies in particular texts are minor, picky, or controversial, his basic point is valid that the accuracy of textbooks for pre-university students is often questionable.

It is my experience that the textbook salesmen mainly push the "readability" of the text, which, indeed, has sometimes been assessed by numerical methods involving counting the percentages of words of different difficulties from an out-of-date list assembled for the purpose. If we can convince school boards and state commissions to concern themselves more with accuracy than with readability, I think students would be better off.

8.9   The value of textbooks

In this World Wide Web-based age, the question is widespread whether students should have a textbook at all. Shouldn't they just use inquiry-based methods, experimenting in class? But can students really figure out by themselves, with whatever guidance, what Kepler, Galileo, Newton, and others accomplished over the years? Current middle-school and elementary-school texts – there are no specific high-school astronomy texts, though college-level survey texts like my own (Pasachoff and Filippenko, 2004) are sometimes used in high schools – contain a mixture of text, labs of varying difficulty and length, notes for teachers, writing activities, mathematical activities, graphing activities, etc. Highly illustrated in color, they can be and are criticized for trying to jam in too much and for providing distracting layouts. Still, the ability of a student to have a reference to a set of things to be learned is invaluable. Further, a large percentage of middle-school and elementary-schoolteachers are untrained in science in general and astronomy in particular, so it makes sense at least to have substantial material available to all in the classroom.

8.10   The ideal textbook

What is the ideal science text? It would have a mixture of activities to intrigue students with scientific methods that would not be so lame or "cookbooky" as to turn them off. It would treat material that the students will find interesting, including modern topics. It would be written by professional scientists, who would have a chance to shape the final version or at least vet it for accuracy – something that is now the purview of editors. It would be thoroughly reviewed by a team of scientists before publication, to make certain that everything that is said is correct. It would be marketed to schools and state adoption committees for its accuracy rather than readability. Most current texts fail in the last two of these desiderata.

But even an ideal textbook may not succeed in the hands of a poorly prepared teacher. Learning is most affected by the interaction of students in a classroom with teachers. So the training of teachers in science is all-important, and relatively few schoolteachers are well trained in science. Salaries and working conditions should improve in order to attract students with better academic records into teaching on the school level. Projects like the Astronomical Society of the Pacific's *ASTRO*, pairing professional astronomers with teachers, are models of teacher training. See http://www.astrosociety.org/education/astro/about/partnerships.html. But how to bring such high-quality training to the majority of schoolteachers remains yet to be determined.

References and further reading

Gagnon, Paul 1995, "What should children learn?" *The Atlantic Monthly*, **276** (December).
Gouguenheim, L., McNally, D., and Percy, J. R. 1998, *New Trends in Astronomy Teaching*, IAU Colloquium 162, Cambridge: Cambridge University Press.
Hurd, Paul DeHart 2000, *Transforming Middle School Science Education*, New York: Teachers College Press.
McClintick, David 2000, "The great American textbook," *Forbes*, October, 178ff.
Partridge, Bruce 2003, "Parting thoughts," *American Astronomical Society Newsletter*, June 2003, 7, 10.
Pasachoff, Jay M. 2000, 2nd edn 2005, *Science Explorer: Astronomy*, Upper Saddle River, NJ: Prentice Hall.
Pasachoff, Jay M. 2002, "What should college students learn? Phases and seasons? Is less more or is less less?" *Astronomy Education Review*, **1**, 124–30.
Pasachoff, Jay M. 2003a, "Rowling got it wrong," letter in *Sky and Telescope*, December.
Pasachoff, Jay M. 2003b, "What should students learn? Stellar magnitudes?" *Astronomy Education Review*, **2**, 4.

Pasachoff, Jay M., and Filippenko, Alex 2004, *The Cosmos: Astronomy in the New Millennium*, Belmont, CA: Brooks/Cole. http://info.brookscole.com/pasachoff.

Pasachoff, Jay M., Pasachoff, Naomi, and Cooney, Timothy 1983a, 1989a (2nd edn), *Physical Science*, Upper Saddle River, NJ: Scott, Foresman and Co.

Pasachoff, Jay M., Pasachoff, Naomi, and Cooney, Timothy 1983b, 1989b (2nd edn), *Earth Science*, Upper Saddle River, NJ: Scott, Foresman and Co.

Pasachoff, Jay M., Pasachoff, Naomi, *et al.* 1989c, *Discover Science*, Upper Saddle River, NJ: Scott, Foresman and Co. (7 volumes, grades K–6).

Pasachoff, Jay M. and Percy, John R., eds. 1990, *The Teaching of Astronomy*, IAU Colloquium 105, Cambridge: Cambridge University Press.

Rowling, J. K. 2003, *Harry Potter and the Order of the Phoenix*, New York: Levine/Scholastic, 718.

Winerip, Michael 2003, "70 per cent failure rate? Try testing the testers," *The New York Times*, June 25, 2003.

Comments

Julieta Fierro: How much freedom do you have writing books for grades K–12?

Jay Pasachoff: Unlike the situation with university textbooks, for which you write basically what you think is appropriate (and which is later reviewed), everything about K–12 textbooks is first laid out by the publisher, especially the space allotment for topics. They may even describe to the authors what is to be on each page or small group of pages. In a recent set of chapters I just wrote, I was given freedom to move things around within a chapter, but there are overlapping and sometimes contradictory state guidelines that have to be met on content.

Bruce Partridge: Earlier, we were urged not to underestimate our students, and that leads to a question: Millions of school children are reading the Harry Potter books, so can you comment on the level of maturity of the prose in Harry Potter as compared to the prose in the K–12 textbooks? Are we, in our textbooks, underestimating our students?

Jay Pasachoff: Anything that gets people reading more – and especially schoolchildren reading more – has the capability of leading students to open and use their textbooks. J. K. Rowling draws people into her books by telling interesting stories, and I hope that the stories of the universe we are telling in astronomy textbooks draw in students to a satisfactory degree.

Case Rijsdijk: Since there are now remote telescopes do you explain to students that there are parts of the sky not visible from the USA?

Jay Pasachoff: Certainly I explain that there are parts of the sky that are always visible from our latitude and corresponding parts that are never visible. A planetarium is useful to enhance that point. I stress the importance of having observatories at low latitudes (like Hawaii's) and in the southern hemisphere (like Chile's and Australia's) in order to see the southern sky objects.

Jay Pasachoff (added in proof): I mentioned Leon Lederman's attempt to interest television networks in a show with an astronomer as a protagonist. A CBS television series in 2004–5 (Numb3rs) has an applied mathematician at Calsci (obviously a Caltech clone) helping his FBI brother solve crimes, using mathematics skills.

9

Distance/Internet astronomy education

David H. McKinnon

Charles Sturt University, Bathurst, New South Wales, Australia

Abstract: Since the early 1990s there has been a proliferation of astronomy courses offered over the Internet. Accompanying the courses has been an increasing number of robotic and remote-control telescopes. Since 1994, Charles Sturt University, Australia, has offered a course on cosmology for gifted and talented high-school students and, since 2000, a remote control telescope for use by elementary- and high-schoolteachers and their students. Both programs are accompanied by extensive resource materials and are offered by distance education to participants. This paper describes many of the outcomes of the research conducted on both projects and what has been learned with respect to the necessary conditions in order that elementary- and high-schoolteachers engage with exciting programs on offer. Professional development of teachers is a key issue if these programs are to be successful.

9.1 Introduction

There have been exciting times since the advent of the World Wide Web in Australia in 1993. The medium has led to a proliferation of courses on the Internet for people interested in a whole host of things ranging from astronomy to astrology, from cosmology to cosmetology, and from celestial mechanics to celestial creativity. Accompanying the courses has been a proliferation of controllable devices: the first robotic telescope was made available in 1993 at the University of Bradford; other projects have been "Telescopes in Education" in 1996, the Charles Sturt University (CSU) Remote Telescope Project in 2000, and the Faulkes Telescope Project in 2003. Other robotic devices in their various forms from cars to manipulators to electron microscopes are also available. Access to these courses and devices is unparalleled – they are available from anywhere in the world to students, teachers, and other educators as well as the interested lay person.

The International Astronomical Union's Commission 46 – Astronomy Education and Development (IAU website, accessed July 7, 2003) takes as its main objective:

to further the development and improvement of astronomical education at all levels throughout the world, through various projects initiated, maintained, and to be developed by the Commission and by disseminating information concerning astronomy education at all levels.

In addition, the IAU's Special Session 4, at the 2003 Sydney General Assembly, of which this volume is the proceedings

deals with K–12 education (primary and secondary level school education). It deals with tertiary and public education only as they apply to K–12 education. The focus is on strategies, programs, and projects which can be shown – by research and evaluation – to be effective in promoting learning of content, skills, applications, and attitudes.

Table 9.1. *Successful completions in cosmology, comparative literature, and philosophy, 1994–2002*

Year	Female	Male	Total cosmology	Total comparative literature	Total philosophy
1994	2	3	5	2	7
1995	2	9	11	10	9
1996	4	9	13	4	23
1997	10	11	21	5	28
1998	6	17	23	10	42
1999	12	15	27	3	49
2000	9	12	21	7	42
2001	8	24	32	7	37
2002	11	9	20	12	55
Totals	64	109	173	60	292

The purpose of this paper, therefore, is to report two major programs being operated by Charles Sturt University, Bathurst, the largest distance education provider in Australia. The two programs are the Cosmology Distinction Course for gifted and talented senior high-school students and the CSU Remote Telescope Project for upper-primary (elementary) school, and junior high-school science, students. The paper concludes with a brief discussion of what is required by projects such as these to have them more widely disseminated and used by teachers in our schools.

9.2   Cosmology Distinction Course

Distinction Courses were first offered in 1994 to gifted and talented high-school students who had completed at least one Higher School Certificate course at the highest course level at least one year ahead of their age cohort and who had been placed in the top 10 per cent of the candidature. Cosmology is one of the three university style distinction courses, with the two others in comparative literature and philosophy. Access is offered to students across the state by distance education no matter whether they live in isolated rural communities or close to a university. Though the number of students enrolling in Distinction Courses is not large, they reflect the number of schools across the state of New South Wales (NSW) that allow students to accelerate in their normal school programs, i.e., not many high schools allow students to progress through their studies faster than their peers. Table 9.1 shows the number of successful participants who completed the courses in the years 1994 to 2002. In 2003, there were 32 students enrolled in the Cosmology Course.

9.2.1   Use of technology to enhance the CDC

In 1994, Cosmology was the first course to employ extensive use of email. In 1995, all students were sent modems to facilitate communication. In 1996, in addition to email, an extensive website was set up by the author with links to resources on the Internet that supported Cosmology and also provided students with access to the latest research findings. The

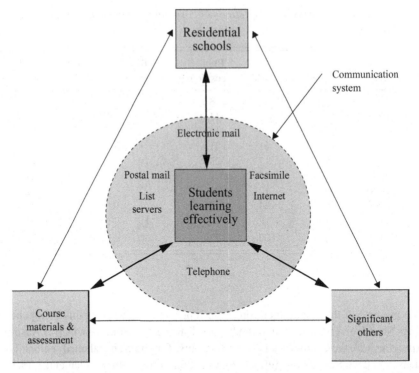

Fig. 9.1. Model of the mixed-mode delivery system (McKinnon and Nolan, 1999).

Cosmology Distinction Course website was incorporated into the NSW Higher School Certificate Online website (2003) hosted by Charles Sturt University.

In 1998, one of the Cosmology students, Carl Gibbs, developed an integrated site comprising a communication forum for cosmology, comparative literature, and philosophy as well as a general communication forum for all three subjects and a resource finder for the latest research in cosmology. The new interactive website came into operation in 1999. Gibbs continues to maintain the site for the New South Wales Board of Studies.

9.2.2   Methodology to evaluate the CDC

A grounded theoretical analysis of the data collected from students' use of the Internet resources and of the electronic communications in the first four years of the Cosmology Distinction Course (1994–7) gave rise to an interactive design model of the way in which the components of the distance education delivery system appeared to work in a successful fashion. The model is presented in Fig. 9.1.

Interactive design model

The model comprises three key design elements: print-based study modules, residential schools, and significant other points, as well as a communication system for linking the students with the elements and with each other. The Course Materials organize and sequence the objectives, content and assessment tasks into manageable units of study. Residential schools bring all of the students together to meet and interact with each other, to engage in experiential learning at world-class observatories and to learn from, and interact with, leading researchers in astronomy and cosmology. Significant other points include the students' peers, the course coordinator, course organizers, research astronomers, and cosmologists who provide students with the support and guidance they may require. During the course the significant others interact with the students in varying capacities as social friends, critical friends, facilitators of learning, mentors, interpreters and discussants.

The communication system connects the students in varying ways with each of the three design elements located at the vertices of the model. The specific means of communication range across such traditional tools as postal mail and telephone to the more modern tools of facsimile, electronic mail, communication forums, ICQ and the Internet. They are the means by which the course coordinator, the students and other key persons associated with the course distribute and receive materials; interactively address and resolve issues, problems, and concerns; share good ideas, communicate their latest discoveries, and interact on a social level with each other unhindered by the "tyranny of distance" in the large state of New South Wales.

The location of students at the center of the model and also at the center of the communication system represented by the shaded circle signifies the student-centered nature of the course and the fact that the locus of control rests with them. Interaction among the design elements creates the environment for students to learn effectively throughout the course. The thin double-headed arrows signify interdependency among the design elements as well as the dynamic and mutually supportive interaction. The thicker arrows indicate that the students interact with each of these elements but that the locus of control rests with them, i.e., when they are studying the print-based course materials, participating actively in residential school activities, communicating with others, initiating, designing and completing a project, and developing new ideas.

Interaction between the design elements and the communication system

The communication system in action mediates all student interactions with the three design elements of the model. It provides them with the means not only to study the contents of the course but also to access a wider range of research information and ideas, and significant other individuals with whom to explore and discuss ideas. The communication system enables this in three main ways: first, through the Internet, and the website especially designed for the course, to access the latest research information in both astronomy and cosmology; second, through the use of electronic mail to communicate personally with peers, the course coordinator and significant others around the state of New South Wales and, indeed around the world; and, third, through the various electronic forums to engage in debates with their peers, and others, on topics related to astronomy and cosmology, or more widely on topics of general interest to young adults with their peers in the other Distinction Courses. One short vignette illustrates the interaction among the course content, assessment requirements, significant others, residential schools and the communication system.

One sixteen-year-old female student joined the course because of her profound interest in astronomy. During the first residential school, while walking back from an observation session on Siding Spring Mountain, she confessed that she wanted to be an "observational astronomer" and that she wanted to do a "practical project." She was encouraged by the course coordinator and by astronomers at the first residential school to pursue her desire. They suggested that perhaps she might join an existing research program being undertaken by a visiting astronomer. Over the next four months, the student made numerous inquiries by electronic mail to check on the progress being made. Enquiries made on her behalf elicited a response from the director of one observatory who arranged for her to meet, and work with, a visiting female astronomer from Italy whose project involved the mapping of dark matter in spiral galaxies visible in the southern hemisphere. The student's project was to be based on a six-day field trip to the observatory. Before undertaking the field trip, the visiting astronomer communicated via email with the student, acting as a significant other, to ensure that the student understood the mathematics required and that she could carry out the relevant calculations correctly. The resulting project produced by the student was awarded a high distinction by the assessment committee. The student graduated from the course and went on to study physics at the University of Sydney. She is now enrolled in the first year of a Ph.D. at the University of Cambridge, UK, undertaking research in astronomy.

9.2.3   Conclusion
The NSW Board of Studies Cosmology Distinction Course, now in its tenth year of being offered by distance education techniques, attracts some of the brightest students in New South Wales. A particular strength of the course is the fact that it can be delivered to students in isolated or rural communities who get access to the latest scientific research in many areas of astronomy and cosmology.

9.3   The Charles Sturt University Remote Telescope Project
Since the early 1990s a small number of robotic and remote telescope facilities have come online and are now accessible over the Internet or through modems. Some of these, the minority, are completely autonomous systems, while others provide remote access in real time to telescopes through the Internet or by direct modem contact. The number of these remote devices has increased dramatically since 2000. In Australia and the UK, the Faulkes Telescope Project involves two 2-m-class robotic telescopes, with one to be located at Siding Spring Observatory, Coonabarabran, and the other in Hawaii. When completed, Australian secondary-school students will have access to the Australian telescope for 15 per cent of the available observing time, while UK schools will have access for the balance. The telescopes will operate both autonomously and under remote control. It is against these large projects that the CSU Remote Telescope Project is framed. The project takes elements from those described above and renders the technicalities of control at a level where primary-age students and their teachers can easily use the system.

The software and hardware systems to drive the CSU Remote Telescope are described elsewhere (McKinnon and Mainwaring, 2000), as are the educational materials written to support the project (McKinnon and Geissinger, 2002). This section of the paper describes the outcomes investigations undertaken to evaluate the impact of the materials on primary and secondary school students' motivation and learning and comments by teachers.

9.3.1   Primary-school package evaluation
The educational materials are supplied in Teachers' Guides entitled *A Journey through Space and Time I and II* (McKinnon, 2001, 2002) and accompanying CD-ROMs. The primary-school guide provides an extensive set of lesson plans covering such topics as: taking control of the telescope and cameras; image processing; finding objects in the sky; the solar system; stars and galaxies; poetry; space travel; constellation myths. The learning materials for primary

schools are integrated across the six content areas of the primary curriculum, while for high schools the materials are targeted directly at science and technology. In the primary school, the materials are designed to engage the students for four hours a day, four days a week for the ten weeks of a school term. For secondary students in grades 7–9, the materials are designed to engage the students for six to eight weeks of normal science periods (approximately six per week). A website also provides an extensive list of resources available on the Internet. When schools come online to take control of the telescope and cameras to image 'their' objects, there is a technician waiting for them to offer help as required.

In 2001, four classes of children in primary Years 5 and 6 learned about astronomy during a ten-week period at four rural schools in NSW. Two of the classes were Year 5 while the other two were Year 5/6 composites. Ages ranged from 9 years to 12 years. The teachers had agreed to participate in the evaluation of the educational materials and in investigating the impact on students' motivation and learning. The impact on students' alternative conceptions was undertaken by a fourth-year honors student as part of her thesis requirements (Danaia, 2001).

Three of the class teachers agreed to undertake the teaching of the materials contained in the Teachers' Guide over a full school term. The fourth class had studied astronomy in the previous school term, while the other three had not had any experience. Thus, the remaining teacher agreed to focus solely on the activities required for the students to understand how to control the telescope, process images and how to find what was "up there" when they were to take control. The teacher's argument for this approach was that "the class had done all of the astronomy anyway." Of the 105 children involved, data for both the pre- and post-tests were obtained from 74 pupils.

9.3.2   Instruments

The tests employed to tap students' tacit knowledge about astronomy were of two kinds: one required students to read questions, indicate whether they agreed, disagreed, or did not know, and to supply evidence for their opinion (Osborne, 1995). The second test required them to draw a picture to illustrate a given situation, such as the phases of the moon, and to provide a written explanation for their diagram (Dunlop, 2000). The two questionnaires acted as both the pre- and post-tests. Thus four scales were generated with good to very good reliabilities (Cronbach alpha range 0.65–0.83), comprising two scales for the correct/incorrect answers and two scales constructed from a Structure of the Observed Learning Outcome (SOLO) analysis of students' reasons (Biggs and Collis, 1982).

9.3.3   Method

A quasi-experimental pretest, post-test design was used to evaluate the impact of the educational materials on students' learning. The pretests that allowed them to express their knowledge about astronomy were followed, in three of the classes, by ten weeks of social constructivist learning across the six key learning areas of the primary curriculum on various aspects of astronomy using the materials supplied in the Teachers' Guide and on the CD-ROM. During weeks 5 and 6, the four classes of students went online and took control of the telescope. The cycle was completed by a post-test that reiterated the questions originally asked. The test results were subjected to statistical analysis using analysis of variance procedures with repeated measures on the occasion of testing. In addition, qualitative data

Fig. 9.2. Case Rijsdijk (South Africa) and David McKinnon (Australia). Photo by Rob Hollow.

were gathered from all participants to demonstrate how the *Journey through Space and Time* program was received and used.

9.3.4   Results

The specific test results are reported in detail in publications by McKinnon and Geissinger (2002) and McKinnon, Geissinger and Danaia (2000, 2003). For the purposes of this paper, only the highlights of the findings are given.

The quantitative results showed that all four classes had only a little general knowledge about astronomy on the pretest occasion with the mean score for all students being 50 per cent of the maximum score. With Dunlop's "draw a picture to explain . . . ," the mean score for all students was 25 per cent of the maximum. This was the case even for the class that had studied astronomy (the *Earth in Space* – on topics related to the solar system) during the first school term. That is, this class had covered the mandatory topic in the Science and Technology K–6 syllabus, visited the library and conducted research on the Internet in order to produce the posters that are a normal demonstration of learning in topics in astronomy in the primary school, yet their general and spatial knowledge was not significantly different from any of the other three classes who had not covered the topic. This says much about this common approach to astronomy in the primary school.

After completing the ten-week unit, students were again tested. Statistical analyses revealed that there was no significant main effect due to the age of the students, class membership, or to gender. There was, however, a significant main effect resulting from the occasion of testing.

That is, the mean of the four scale scores for all of the students was significantly higher on the post-test than on the pretest occasion (effect sizes ranged from 0.5 to 0.75). On average, students showed a significant improvement in both their general and spatial knowledge. Similarly, the sophistication of their explanations improved significantly showing that they had learned how to express their ideas better in both written and pictorial form. There was also an interesting significant class × occasions' interaction for each of the two instruments. In general knowledge, three of the classes showed a significant learning effect. The teacher of the fourth class had decided to concentrate the class's efforts on the solar system; these children's general knowledge did not rise significantly. The one class with prior experience of astronomy showed a significant rise in knowledge.

With respect to the spatial knowledge instrument, again three of the classes showed a significant learning effect. In this case, the score of the class with prior experience of astronomy did not rise at all. It should be remembered that this particular group did not engage with the learning materials devoted to concepts about the solar system, including orbits, day and night, seasons, and phases of the moon, but instead concentrated on the technical aspects, such as telescope control and image-processing procedures. On average, the students in this class could only achieve a score of 20 per cent.

One aspect clearly demonstrated by this research was that the class that concentrated mainly on the technical aspects showed little change in its alternative conceptions of solar system phenomena, yet this had been the topic of their previous term's work in the *Earth in Space* unit. Students in the other three groups demonstrated that a large number of them had substituted more scientific explanations for their previously held alternative conceptions.

The researcher believes that the students' explorations of various astronomical phenomena, coupled with peer discussion and verbal reworking of concepts, enabled them to discard some of their more naive ideas. The technology that allows children to take control of a sophisticated telescope via the Internet is a great motivator, helping sustain their interest and thus allowing them to engage with the content and learn about astronomical objects and their relative positions in space. The evidence supports the position that engaging with such concepts leads to more insightful learning than does a concentration on mere technical details of controlling a telescope. Another powerful motivating factor was the children's ability to take actual photographs, download them to their school computers, process the images, and display them for peers, parents, and the general school community. The children were deeply interested in what they could see with the telescope and what they could show to demonstrate their new knowledge and skills.

In summary, the *Journey through Space and Time* program in primary years (grades) 5 and 6 has demonstrated good immediate learning outcomes. It remains to be seen whether the students will maintain their interest in the stars and add to their scientific and astronomical knowledge in the future. Their familiarity with the Internet as a source of up-to-date astronomy information plus their new-found ability to conduct email discussions with peers and experts will stand them in good stead.

9.3.5   High-school package evaluation: Canada and the Netherlands

The secondary educational materials are supplied in Teachers' Guides entitled *A Journey through Space and Time II* (McKinnon, 2002) and a CD-ROM. The high-school guide also provides an extensive set of lesson plans covering some similar topics to the primary package such as: taking control of the telescope and cameras; image processing; finding

Table 9.2. *Planned observation run by 4 VWO, St.-Canisius, Almelo, the Netherlands*

Name of the object	Constellation	Type of object	Exposure time
1. NGC 1232	Eridanus	Spiral galaxy	??

We want to look at NGC 1232 because it is a beautiful example of a spiral galaxy. It looks like our milky way.

| 2. Saturn | Taurus | planet | 0.05–0.08 seconds |

We want to look at Saturn because the rings can't be seen with the unaided eye. We have a question: Do the rings spin solitary from the planet? The rings of Saturn are made of what?

| 3. Eskimo nebula | Gemini | constellation | ?? |

We have a question: is it possible to take a picture of the Eskimo nebula? We saw this nebula on picture from the internet.

| 4. Jupiter | Cancer | planet | 0.001–0.006 seconds |

We want to look at Jupiter to see it more nearby. It is a planet just like Earth and with the telescope we can see Jupiter very well. Maybe we can see differences with Earth. We have a question: Could you tell us something about the planet Jupiter? Is Jupiter having rings?

| 5. M68 | Corvus | Globular cluster | ?? |

We want to look at M68 to do something else. There is no group who is imaging a globular cluster. We have a question: What is a globular cluster?

objects in the sky; the solar system; stars and galaxies. It does, however, focus more on the scientific aspects of measurement, experimentation and data reduction, for example, measuring stellar magnitudes, distance determinations, and experimenting with models involving lunar cratering.

Two groups of junior secondary-school students from the Netherlands and Canada participated in the evaluation of the materials for junior secondary science during 2003. A total of approximately 350 students supplied data. The Dutch data have yet to be fully analyzed in collaboration with the University of Twente, Enschede. In this section, qualitative data is provided to illustrate the fact that science teachers in both countries are highly impressed with the curriculum materials, the motivation of their students and the ease with which the system can be used from the other side of the world. One Dutch teacher supplied the following in an email the day after the class's online session where they had planned for and taken images of ten objects on the night of February 8, 2003. She wrote:

In my opinion the session was a great success! The students were very pleased by the fact that taking pictures was very easy to do and they were surprised that there was someone sitting at the other side of the world to help them. When you were talking to them through Notepad or took over the mouse to help them, some students were astonished to see that. They also felt sorry for you, that you have to help them in the middle of the night. In general: My experience is very positive. Next year (if that is possible) I will do this assignment again. I hope that your experience was positive also, so that we can use your telescope again.

Each class provides additional data when they submit their observation proposal and book telescope time. One Dutch class supplied the extract of data shown in Table 9.2. The proposal list indicates the level of naivety that exists in grade 9 students.

One teacher in Manitoba, Canada, finds it impossible to do any practical work in astronomy during the winter months when the temperature commonly goes down to −40 °C. During 2002–3, the teacher ran the secondary program with three grade 9 classes and also with the school Astronomical Society, which undertook a number of projects including mosaics of extended objects such as the moon, M42, and Omega Centauri, and multiple exposures of faint objects to practice stacking the images. The classes collaborated to take exposures of Uranus, Neptune, and Pluto, and traded their images to see how these planets could be identified by their movement against the starry background. After a session with the Astronomical Society, the teacher sent this email:

Thank you once again for the help and advice during the session. The students were impressed with the individual images and we haven't even had a chance to enhance or stitch!!! It was actually the students and not myself who were experimenting with the exposure time (you could tell from the spelling!!). They learned a lot doing this way. I didn't get to do anything this time around so we have certainly elevated their confidence over the last two sessions. I'll let you know how the mosaicing goes and I'll contact you about July 3 in a couple of days. Thanks again.

This same teacher also claimed later, following a session online with a grade 9 class containing "at risk" students, that:

From an educational point of view ... when "at risk and beyond" kids produce a report with enhanced images of their choice accompanied by a bit of research I know we have done something very significant over the past few months. This is priceless!!

There is little doubt that taking control of a telescope over the Internet is a motivating experience for both primary- and secondary-school students. The control aspect sustains their interest in the science over a considerable period of time. It is, perhaps, this capacity of the control dimension to ignite and sustain interest that allows teachers the unusual luxury of addressing the deeper scientific content and students' alternative scientific conceptions. There are, nonetheless, many issues that need to be addressed, some of which are dealt with in the discussion.

9.4   Discussion

What has been learned from the Cosmology Distinction Course and the CSU Remote Telescope Project offered by the university? It is clear that in the cosmology course we are dealing with an unusual group of students far removed from the mainstream of normal secondary education. They are committed, intelligent, motivated, persistent, and become passionate about science. In short, the students appear to possess many of the attributes that we would wish to see in undergraduate students in our science courses at university and later as researchers in whatever scientific field they choose to enter. What caused them to be interested enough in science to undertake the challenging task of studying cosmology by distance education may in part be due to their early experiences in science or experiences in secondary school, both positive and negative, but nonetheless remains an open and individual question. What we do know is that most of them are bored by science as it is taught in most of their high schools. The key question to answer here is what does the Cosmology Distinction Course offer them that their schools cannot? Part of the answer at least is clear. The students get access to committed, intelligent, motivated, persistent and passionate researchers in astronomy and cosmology who are excellent communicators. It was interesting again to note the reaction of the students at their last residential school held in June 2003, after they had listened to, and interacted with, the speakers: Dr. Joss Bland-Hawthorn on the topic of "Galactic winds;" Dr. Zdenka Kuncic on "Supermassive black holes and active galaxies in the universe;" Dr. Maria Hunt on "Organic molecules and life in the universe;" and, Dr. Charlie Lineweaver on "The cosmic microwave background and the early universe." The students' interest was palpable. Their questions flowed over into the coffee breaks, and later continued in conversations on the Internet forum and with the course coordinator as they finalized topics for their major research projects. This is a situation all too rare for the vast majority of students, who could become interested in science but do not for a variety of reasons.

A second key aspect of the course is the support that they get, which is available through the digital communication channels as they study the content. The students are not simply abandoned to sink or swim. Indeed, few of the students choose to withdraw from the course. The completion rate is extremely high. In most years it is greater than 90 per cent. Science delivered in this way by distance methods has to be "high touch" (Naisbitt and Aburdene, 1990). The students are mentored during their candidacy and later when they go to university by the alumni of earlier cohorts. They feel a part of a community of learners.

The case for the CSU Remote Telescope Project is rather similar. The fact that the telescope is not a robotic device has many advantages (and many disadvantages). The most obvious advantage is that it is "high touch." This in fact is the major lesson to be learned from this project. Primary teachers and secondary science teachers are, in the main, not adventurous. Both sets appear to need their hands held while they prepare to use the telescope, while their students use the telescope, and later in the debriefing sessions as their students process the images.

In a recent professional development online session with science teachers in Manitoba, the issue of hand holding was evident in the evaluation paper sent to me by the organizer:

Everyone got to try the software and I had several comments on how easy it was to get through it. At first I think they were intimidated (or overwhelmed) but it certainly took off.

It would seem that if we wish to break the fear barrier and encourage teachers to take control of the online telescopes, whether they are remotely controlled or robotic devices, then we need a development program that covers a number of years. The author has chosen to start this program in the upper levels of primary school and to carry this forward into the junior secondary area. Given the "fear factor," or perhaps it is a "technophobic factor," the "high touch" environment of the remote control devices helps teachers gain confidence.

The author hypothesizes that, if an evolutionary and developmental approach is taken to the introduction of these devices, there will come a time when science teachers will be prepared to add remote and robotic telescopes to their armory of technology in the secondary science laboratory. At present, it is the high-risk takers who are prepared to employ these devices. Later, the less enterprising individuals will decide to adopt and implement the innovation if it can be demonstrated that there are tangible benefits. This is the way educational innovation normally proceeds. In many respects, this is not surprising. What is surprising is that researchers in the science education field are not treating the introduction of these devices in ways that educators would treat the introduction of computer technology, team teaching, or a new mathematics syllabus. That is, remote and robotic telescopes should be treated as an educational innovation that has to satisfy the criteria for adoption and implementation, and, further, that the introduction of an innovation into education is a long, difficult, and tortuous process (Adams and Chen, 1981; Barnett *et al.*, 1999). The lesson to be learnt from the many other educational innovations that have not satisfied these criteria is that they do not happen.

References

Adams R. S. and Chen, D. 1981, *The Process of Educational Innovation: An International Perspective*, London: Kogan Page.

Barnett, B. G., Hall, G. E., Berg, J. H., and Camarena, M. M. 1999, "A typology of partnerships for promoting innovation," *Journal of School Leadership*, **9**(6), 484–510.

Biggs, J. B. and Collis, K. F. 1982, "Evaluating the quality of learning: the SOLO taxonomy," in *Structure of the Observed Learning Outcome*, New York: Academic Press.

Danaia, L. 2001, "Students' alternative scientific conceptions: an intervention involving the Charles Sturt University Remote Telescope," Unpublished B.Ed. (honors) thesis, Charles Sturt University.

Dunlop, J. 2000, "How children observe the universe," *Publications of the Astronomical Society of Australia,* **17**, 194–206, http://www.atnf.csiro.au/pasa/17_2/dunlop/

Hollow, R. 2003, "Engaging gifted students through astronomy," Paper presented at Special Session 4, Effective Teaching and Learning of Astronomy, XXV General Assembly, IAU, Sydney, Australia, July 13–26.

McKinnon, D. H. 2001, *A Journey though Space and Time I: Teachers' Guide*, Bathurst: Charles Sturt University.

McKinnon, D. H. 2002, *A Journey though Space and Time II: Teachers' Guide*, Bathurst: Charles Sturt University.

McKinnon, D. H. and Geissinger, H. 2002, "Interactive astronomy in elementary schools," *Journal of International Forum of Educational Technology and Society*, **5**(1), 124–8.

McKinnon, D. H. and Mainwaring, A. 2000, "The Charles Sturt University Remote Telescope Project: astronomy for primary school students," *Publications of the Astronomical Society of Australia*, **17**(2), 133–40.

McKinnon, D. H. and Nolan, C. J. P. 1999, "Distance education for the gifted and talented: an interactive design model," *The Roeper Review*, **21**(4), 320–5.

McKinnon, D. H., Geissinger, H., and Danaia, L. 2002, "Helping them understand: Astronomy for Grades 5 and 6," *Information Technology in Childhood Education Annual 2002*, **1**, Norfolk, VA: Association for the Advancement of Computers in Education, 263–75.

McKinnon, D. H., Geissinger, H., and Danaia, L. 2003, "Learning about astronomy: the Remote Telescope Project in primary schools," in Chandra B. Sharma, ed., *Technology Enhanced Primary Education: Global Experiences*, New Delhi: Kautilya Publications.

Naisbitt, J. and Aburdene, P. 1990, *Megatrends 2000: Ten New Directions for the 1990s.* New York: Morrow.

Osborne, J. 1995, *Common Ideas in Astronomy Questionnaire.* London: Kings College.

Further reading

IAU Commision 46 website, http://physics.open.ac.uk/IAU46/, accessed July 7, 2003.

NSW HSC Online is at http://hsc.csu.edu.au/ and now covers 45 HSC subjects. Cosmology was the first subject to go online in June 1996.

Comments

Paul Murdin: In showing the list of competencies that can come from astronomy teaching, the strength of astronomy as a teaching tool came across to me, as politicians and civil servants perceived it when I worked in UK government departments and ministries. They were not really interested in whether students learned about the solar system or the universe, although they appreciated that these were interesting topics that motivated students. They saw astronomical investigations as models for the way the world operates. You see something, you study it, you fit it in to a context, you make measurements, tabulate, graph, display and interpret, and draw conclusions, communicating all this to others. What was earlier described as a disadvantage of astronomy – that it is outside the paradigm of experimental science – was seen by the government as an advantage, because it is like the business world operating in the marketplace, or like any aspect of human life, in which you make up your mind about something after some sort of investigation based on incomplete data. Government saw the teaching of astronomy as a way to teach an attitude rather than facts, with the astronomical facts of very little economic value, but the astronomical methodology as a way, to put it bluntly, to develop the economy of the UK by creating people trained in understanding what they see.

Guido Chincarini: The talk has been extremely interesting – what I did not get, however, is the content of the cosmology module and the way you communicate the concepts to a 14–15-year-old student, for instance. For example, how would you communicate the concept of recombination or similar matters?

Lars Lindberg Christensen: Magnificent results! How can we benefit from your experiences? Can we have access to your cosmology teaching materials? Have you published your results?

David McKinnon: Yes, you can have access, and yes, I have published them. There are quite a lot of papers on these topics.

Anonymous: Did your students ever work in groups in the cosmology course?

David McKinnon: No, though students did collaborate over the Internet. Our education system is still individualistic and competitive, unlike the American system, which is more collaborative.

Julieta Fierro: How many students can you handle in your programs and how do you select them?

David McKinnon: One remote telescope at the site in Bathurst, where there are approximately 210 clear nights per year, can potentially handle a maximum of 1,800 observing hours. In order to ensure that a class of students (approximately 25–30 pupils) gets a scheduled session on one of the three nights that they "request" when the objects they wish to observe are visible, we limit the use of each telescope to 1,000 hours per year. This means that 1,000 classes of primary and secondary students, each of 25–30 students, can use the system. The potential maximum audience for one remote control telescope is thus 1,000 teachers and 25,000–30,000 students around the world. This is a significant potential target population. Selection of classes is not handled by Charles Sturt University. Teachers normally make the first contact and request to be a part of the project. When they decide that they will engage with the project, educational materials are sent to them and their class. They assume the locus of control and request time when it suits them. An online booking system shows them how their request fits in with the usage of the telescope so that they can make changes as necessary.

Open discussion

Educating students using robotic telescopes

Case Rijsdijk: Robotic telescopes are an oxymoron – how much time in terms of human resources is needed to maintain a telescope?

Jayant Narlikar: Our center, the Inter-University Center for Astronomy and Astrophysics (IUCAA), in Pune, India, uses the Internet to operate a small telescope at Mt. Wilson, California, for school children at Pune. There is a time difference of $12\frac{1}{2}$ hours, which makes it possible for the schoolchildren to use direct observing methods. I think all such groups that use the Internet for remote observing by schoolchildren should get together and exchange their experiences. It would help to have an email directory of such groups.

Nick Lomb: Remote-controlled telescopes are an important new teaching resource. However, we need to make observation with them as exciting as possible. We suggest real-time observing with contact with an observer at the telescope. Adding "bells and whistles" such as a webcam showing the motor of the telescope would be most useful. A second problem is overcoming teacher reluctance to try new technology as well as training them to have enough astronomical background and technical knowledge to operate the telescope.

Jay Pasachoff: If someone, perhaps one of the speakers, wants to make a list of remotely activated telescopes – and keep the list current – we would be glad to link to that list from the website of our IAU Commission 46 on Education and Development. It strikes me that these remote telescopes might be a very useful way to bring astronomical observing to developing countries.

John Broadfoot: Some comments:

(1) I agree wholeheartedly with McKinnon and Tsubota that students need to use a "real" telescope;
(2) Telescopes In Education Project – there are problems to resolve;
(3) Trialed at Queensland University of Technology – use of a webcam and a CCD camera connected to a laptop provides a useful compromise to enable groups to view images.

David McKinnon: I think it's tremendously important for students to look directly in the eyepiece of a telescope.

Harry Shipman: The last speaker expressed a need for wider coverage of the sky. There are several all-sky cameras available on the Internet at www.nightskylive.net. I have used them successfully for student activities.

Geoffrey Wyatt: Please don't ignore the average, mediocre students in favor of the gifted or talented few. Also, don't foster the "black box" image of robotic telescopes. Make them visible by webcams, or through an online chat with a controller.

Yukimasa Tsubota: We can use the remote telescopes to excite the students, but it is not easy to use them within the constraints of the national curriculum. We need to develop good lesson plans for using remote telescopes. I recommend implementing a network of wide-angle cameras all over the world.

M. Dennefeld: The previous speakers have concentrated on the use of robotic telescopes for students but, as McKinnon has mentioned, the problem is that the majority of teachers are scared by the technology. Are there any plans to use the robotic telescopes to teach the teachers themselves? Note that professional astronomers now have plans to use some time on the existing 1–2-m research telescopes to organize schools for teachers as the telescopes become somewhat more accessible. The pressure of research has shifted from these smaller telescopes to the 8-m telescopes.

Harry Shipman: In Delaware, astronomy is one of the major strands in the state curriculum; however, there are other places it comes in. For instance, a team of teachers and scientists, which I led, developed a six-week energy curriculum unit. One topic to cover was the electro-magnetic spectrum. Rather than do the physics first, we used infrared astronomy to introduce the electromagnetic spectrum. This way of giving astronomy a place in the curriculum – astronomy by stealth, as it were – may not show up as a four-week unit.

Jay Pasachoff: There's a lot of good curriculum material developed by SIRTF (now Spitzer Space Telescope; www.spitzer.caltech.edu), including a ten-minute video.

Harry Shipman: In Delaware we use that video.

Poster highlights

Some of the varied astronomy teaching methods are examined here, starting with **Paul J. Francis**'s paper, **Using games to teach astronomy**.

I have been experimenting with using role-playing games to teach introductory university astronomy. The idea is this: rather than simply telling students about some topic (e.g., the climate of Venus), I tell the class to "imagine that you are world experts on Venus, gathered together here at great expense to solve the baffling mystery – why is Venus so much hotter than the Earth?" The class is divided into small groups, and each group is given a briefing paper. A group, for example, might be experts on infrared radiation, or atmospheric transparency, with their briefing paper giving them a set of clues on this topic (along with lots of red herrings – to teach students the art of extracting meaningful information from noise).

No single briefing paper contains enough information to solve the puzzle – students have to wander around the room, exchanging clues, and slowly putting together a plausible theory, which they then present to the rest of the class.

How does it work? Fabulously well, in general. It really gets students thinking, and interacting with each other. It permanently changes the whole classroom dynamic. At first there was concern that students would go berserk (and a security guard once tried to close down one of these lectures, thinking it was a riot in progress), but even poorly motivated high-school students seem to find these exercises interesting enough to keep their attention. The only major problem is that occasional classes get overly competitive and start lying to each other (behavior seen so far in one prestigious US university and in an Australian primary school).

More details, and copies of lots of these exercises can be found online at: http://msowww.anu.edu.au/~pfrancis/roleplay.html

Comments

Michael Drinkwater: I have used your role-playing exercises with enjoyment in university teaching. Have you any comments on their use for secondary- or primary-school teaching?

Paul Francis: They work well at school level if the appropriate exercises are used.

Ronald G. Samec's innovative Mission to Mars program is explained here as a way to get students excited about astronomy in **Attracting students to space science fields: mission to Mars.**

Attracting high-school students to space science is one of the main goals of Bob Jones University's annual Mission to Mars (MTM). MTM develops interest in space exploration

119

Paul Francis (Australia) interacting with teachers at the July 26, 2003, teachers' workshop in Sydney. Photo by Rob Hollow.

through a highly realistic, simulated trip to Mars. Students study and learn to appreciate the challenges of space travel including propulsion, life support, medicine, planetary astronomy, psychology, robotics, and communication.

Broken into teams (Management, Spacecraft Design, Communications, Life Support, Navigation, Robotics, and Science), the students address the problems specific to each aspect of the mission. Teams also learn to interact and recognize that a successful mission requires cooperation. Coordinated by the Management Team, the students build a spacecraft and associated apparatus, connect computers and communications equipment, train astronauts on the mission simulator, and program a Pathfinder-type robot.

On the big day, the astronauts enter the spacecraft as Mission Control gets ready to support them through the expected and unexpected steps of their mission. Aided by teamwork, the astronauts must land on Mars, perform their scientific mission on a simulated surface of Mars, and return home. We see the success of MTM not only in successful missions but also in the students who come back year after year for another MTM.

The process of using astronomy as a hook for gifted students is described next in **The Cosmology Distinction Course for gifted students in New South Wales** by **Robert Hollow, D. McKinnon, C. Gibbs, R. Holmes, and B. McAdam.**

The Cosmology Distinction Course is a one-year course offered to gifted senior high-school students in New South Wales (NSW), Australia. Currently in its tenth year of operation, the course is delivered by distance education methods and residential sessions. Students visit key astronomical observatories in NSW, meet, and interact with, researchers in cosmology and astronomy. Course assessment is by assignments, external exams and a major project. It contributes to students' matriculation results for university entry. A website developed by one of the students allows students and lecturers to communicate with each other and access resources.

References (not comprehensive)

Hollow, R. P. *et al.* 1994, "The Cosmology Distinction Course, NSW," *Proceedings of the Astronomical Society of Australia*, **11**(1), 39–43.

Hollow, R. P. 1995, "The Cosmology Distinction Course for gifted students," *Physics Education* **30**(3), 129–34.

McAdam, W. B. 2000, "Access to astrophysical research by secondary students," *Publications of the Astronomical Society of Australia*, **17**, 163–70.

A mentorship program, in which high-school students are paired with professionals working for a leading research institute in Texas, is described in **Young engineers and scientists: a mentorship program** by **Daniel C. Boice**.

The Young Engineers and Scientists (YES) Program is a community partnership between Southwest Research Institute (SwRI), and local high schools in San Antonio, Texas (USA). It provides talented high-school juniors and seniors a bridge between classroom instructions and real-world, research experiences in physical sciences and engineering.

YES consists of two parts:

(1) an intensive three-week summer workshop held at SwRI, where students experience the research environment first-hand; develop skills and acquire tools for solving scientific problems; attend mini-courses and seminars on electronics, computers and the Internet, careers, science ethics, and other topics; and select individual research projects to be completed during the academic year

(2) a collegial mentorship where students complete individual research projects under the guidance of their mentors during the academic year and earn honors credit.

At the end of the year, students publicly display their work, acknowledging their accomplishments and spreading career awareness to other students and teachers.

YES has been highly successful during the past ten years. All YES graduates have entered college, several have worked for SwRI, and three scientific publications have resulted. Student evaluations indicate the effectiveness of YES on their academic preparation and choice of college majors.

A poster on updating astronomy exercises for online use is provided by **Richard Gelderman** in **Java-based exercises and tutorials – updating the classics**.

A great deal of recent emphasis has been focused on hands-on, interactive lessons that place the learning into the hands of students. Not that this is a new idea; in fact, generations of secondary students and current K–12 educators were trained with such classic activities as "Laboratory exercises in astronomy" (*Sky and Telescope*) or "Astronomy through practical investigations" (LSW Publications).

Regardless of their high-quality content, however, these stalwart pen-and-paper activities simply are not in synch with the current generation. Luckily, the proliferation of Internet-connected computers provides the opportunity to revise previously successful activities for today's students. We have created Internet-deliverable versions of such familiar activities as "Stellar spectral classification," "Galaxy classification," and "Stars, gas, and dust in the Milky Way," designed for use as both ungraded tutorials and exercises assigned for a grade. To ensure the activities can be accessed by the largest audience, they have been written as Java-based scripts compatible with all web browsers. We shall present demonstrations of completed activities, an evaluation of our experience, and plans for the future.

Editor's Note: See also Larry Marschall's *Project CLEA* (Contemporary Lab Exercises in Astronomy), computer labs: www.gettysburg.edu/academics/physics/clea/CLEAhome.html.

Further high-tech teaching methods exclusively for introductory college-level courses are explored in **Innovative technology for teaching introductory astronomy** by **Mike Guidry**.

The application of state-of-the-art technology (primarily Java and Flash MX Actionscript on the client side and Java, PHP, PERL, XML, and SQL data-basing on the server side) to the teaching of introductory astronomy built around more than 350 interactive animations called "Online journey through astronomy," and a new set of 20 online virtual laboratories are available, for example, with the textbook *The Cosmos: Astronomy in the New Millennium* by Jay M. Pasachoff and Alex Filippenko (at info.brookscole.com/pasachoff). In addition to demonstration of the technology, our experience using these technologies to teach introductory astronomy to thousands of students in settings ranging from traditional classrooms to full distance learning are summarized. Recent experiments using Java and vector graphics programming of hand-held devices (personal digital assistants and cell phones) with wireless wide-area connectivity for applications in astronomy education are described.

The electronic astronomy that we have developed has not been evaluated for effectiveness in a formal way, but we have used it in various guises with several thousand students in introductory astronomy at the University of Tennessee over several years. That has given extensive anecdotal experience indicating that the approach is very effective for a particular subset of students (those who can manage their own time and who appreciate flexibility in scheduling). It is less effective with the students who are there merely for a grade and who cannot manage their own time. The latter students don't seem to do any more poorly than their counterparts in a regular section, but they don't do any better either.

Two observations:

- A steady decline in the enrollment for introductory astronomy reversed when we introduced our first extensive electronic material five years ago; since that time we have routinely had more applicants for introductory laboratory astronomy for non-science majors than we can accommodate (our lab space limits us to about 600 students per semester in astronomy courses requiring a lab).
- We have taught an optional course for three years that uses only the electronic syllabus (no textbook), and has no formal lecture (approximately a distance learning environment for lecture, but specifically for on-campus students). The students have a lab in which half of the labs are virtual and done on their own schedule, but half are "real" and done in the laboratory with a teaching assistant. In comparing that course with standard lecture with textbook sections of intro astronomy that I have taught in the past, I found that the grades were generally higher. For example, the first semester that I tried this I had about 180 students in initial enrollment for the Fall semester and ended up giving 25 per cent of

students completing the course an A. In a normal lecture course for the same material I have seldom given more than 10–15 per cent A grades. A subjective comparison of test questions indicated that the tests for students in the electronic sections were comparable to or somewhat harder than those given by professors in other sections of the same course (who used a textbook and had formal lecture), and also comparable to or somewhat harder than tests I have given in normal lecture sections of the same course. These results are highly suggestive but are of course still subjective, since the test and control students did not take exactly the same tests.

Expounding further on high-tech applications for teaching astronomy, **Masae Muraoka** gives us a poster called **Virtual reality as a tool for astronomy education**.

Although astronomy is one of the most popular subjects in education, we are continually encountering difficulties in providing young students with observational experiences, because astronomical phenomena need to be observed mostly at night. To avoid such difficulties, the authors present virtual reality as a workable alternative which frees educators from observational constraints (i.e., weather conditions, security).

As an example of using virtual reality in the curriculum, they created a 3D stereographic simulator of the Leonid meteor shower. Using 3D data obtained in 2001, they created software through which students can experience the meteor shower with active stereographic images. The active graphics provide images from almost any viewpoint. Students can observe not only from anywhere on Earth but also from the center of the shower (about 90 km from the Earth's surface). They are also able to zoom out into outer space and play the movement back at any speed. When students approach the meteor they can find detailed structure of the shock and the coma.

As an additional function, they can also simulate the Leonid meteor shower using number count data and enjoy its peak at any time.

A program from France that gets students to make their own measurements to determine the size of the Earth is described in **Measuring the circumference of the Earth at primary school** by **Emmanuel Di Folco**.

The French program "La main à la pâte" is leading an international and cooperative project gathering each year more than 100 schools all over the world. "Following in the footsteps of Eratosthenes" invites teachers and pupils at primary schools to measure the circumference of the Earth following the method first developed by the Greek scientist Eratosthenes 2,200 years ago.

The protocol consists of a series of experimental activities that allow a progressive approach to the various scientific notions at play (light rays and shadows, the shape of the Earth, the solar noon, etc.). Pupils are invited to reproduce the observations of Eratosthenes and to adapt his method by developing their own instruments. Finally they can easily compute their own estimation of the size of our planet by exchanging their measurements through the Internet with other classes from many countries.

We will present the protocol and its activities as well as the specific cooperative tools that have been created to help and follow up the teachers in the course of the project: a scientists' and trainers' network, an Internet forum and a data base where all the measurements gathered can be exchanged among the participants.

Comments

Jay Pasachoff: For the Eratosthenes project, how do students find out when local noon occurs, given that it can be very different from noon on a clock?

Response: They measure it from the time of minimum shadow.

The Internet can provide global access to witnessing astronomical events, as is explained in **Astro-classroom for high school students of the world** by Japan's **Jun-ichi Watanabe**.

"Astro-classroom for high school students of the world" was established in 1998, mainly in anticipation of the expected strong activity of the Leonid meteor shower over Japan; 248 high schools participated. Since then, we have continued this campaign to let high-school students watch genuine starlit skies to experience an astronomical phenomenon with their own eyes, and to study it themselves. The target astronomical phenomena included the Leonids in 1999, 2000, 2001, and the lunar eclipse of 2000. From 2001, we selected two or three phenomena in a year such as an occultation of Jupiter, the Perseids, and a solar eclipse. With the help of the Internet, the system is easy for school students, who need only register themselves at the webpage of our campaign, and report their results after making the observations.

The students prepare detailed manuals explaining how to observe on their webpage. More than 5,000 students have experienced real astronomical observations by participating in this campaign. Actual scientific results have been also obtained from these activities on the Leonids. We introduce our activities in this paper. Detailed information is given at http://www.astro-hs.net/.

Another representative from Japan, **Norio Okamura**, recounts an experience working with high-school students to create a **Reproduction of Cassini's telescope**.

As an activity of the astronomical club in the Mito Second High School, we tried to reproduce a classical telescope similar to that used by Giovanni Domenica Cassini. He discovered the Cassini gap in the rings of Saturn by using a 10-cm telescope with a focal length of 11 m. We asked an optical engineer, Mr. Hidaka, to make a lens. We then brought the students of our astronomical club to the factory, and let them polish it, following his suggestions. The eyepiece was also made from a combination of two single lenses. The frame of the supporting telescope was made first from wood, along with some iris masks. However, the wood frame was too heavy to handle, so we replaced the wood frame with a metal frame with wood mounting, and succeeded in taking an image of Saturn using this telescope (see figure on page 125).

All the fabrication processes over a two-year period were carried out by the students themselves. Making real reproductions at such a large scale seems to be a wonderful experience for the students.

In 2003, students and others have observed with this model of Cassini's telescope a total of seven times. About 300 people have watched the moon and Mars through the telescope. Students in my club gave a lecture on the telescope and showed other people how to use it (see figure on page 126).

Yukimasa Tsubota, also from Japan, recounts the experience of **Teaching astronomy and telescope use in a high school**.

Many people view the art of astronomy as observing the stars through a telescope. Many schools have small telescopes, but it's not easy to use them in science classes. For instance, only the sun is available during class time, and many teachers don't know how to use the

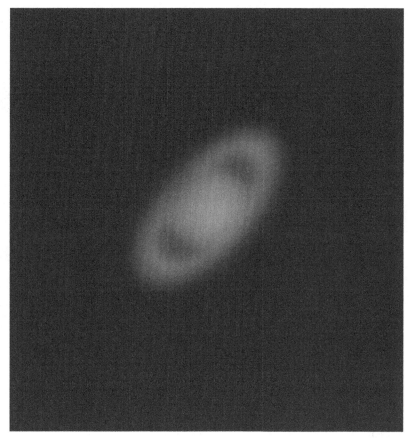

A picture of Saturn captured with a reproduction of Cassini's telescope with which he discovered the Cassini gap in the rings of Saturn (see article by Okamura, p. 124).

telescopes safely to observe it; in addition, there is only one telescope per 30 students in a class. Suppose we have a remote telescope that enables us to see the stars at night from elsewhere on our planet while the daytime class is in session? That is possible via the Internet Astronomical Observatory, developed at Keio Senior High School, which is available for the student at any time, from any place.

Real-time images of the moon and planets impress students. However, the remote telescope would be used more effectively within the school's science curriculum under the tutelage of a teacher, since the images aren't as good as the images produced by the bigger observatories. Real-time celestial observation is as entertaining as it is educational, hence students are motivated to learn astronomy. Using the Internet Astronomical Observatory requires the development of a telescope-use plan. Keio's lesson plan is also discussed here. They also invite the reader to experience the Internet Astronomical Observatory.

The classic approach to viewing the sun through a pinhole is expanded upon here by **Costantino Sigismondi** in **Solar astronomy with pinholes**.

A flat mirror projects sunlight onto a framed pinhole. The pinhole produces the solar disk image of diameter D_i on a screen parallel to the frame posed at focal distance f, better in a

Students with a reproduction of Cassini's telescope.

room. For an ideal point-like pinhole, the angular solar diameter is $\tan(D) = D_i/f$, and can be measured:

(1) like Kepler, comparing the disk with pre-drawn disks of known diameter;

(2) measuring D_i on the image (possible errors: identifying the true diameter among chords; problems due to the motion of the image; limb darkening);

(3) timing the passage of the disk perpendicularly to a profile posed before the screen (possible errors: not a perpendicular path; uncertainty on the limb; contact times);

(4) using two equal pinholes built on the same frame at distance d between centers, measuring the focal length, f_c, where the disks are in contact; $\tan(D) = d/f_c$ (possible errors: f_c uncertainty).

All methods have to deal with systematic errors due to diffraction and to the finite opening of the pinhole. The data analysis is discussed in the four methods. They are experiments with easy-to-find and low-cost material, approved for indoor demonstrations, and offering students different levels of complexity in setup strategies and data analysis.

Next, the situation of astronomy education in Queensland, Australia, is explained in **Astronomy education in Queensland schools** by **Stephen Hughes** *et al.*

The NTQO (Nanango TIE Queensland University of Technology (QUT) Observatory), about 200 km north-west of Brisbane, Australia, opened in October 2002, and comprises a C14 telescope on a robotic mount (Paramount, Software Bisque) with an Apogee AP7b

512×512 CCD camera. A second C14 is in the process of being installed, and the observatory can take up to five telescopes. The observatory is linked to the outside world by a satellite link with a download speed of 512 kb/s and an upload speed of 128 kb/s. The observatory is powered by an array of 12 solar panels generating 5.2 kW h per day, feeding a bank of batteries that can power the observatory continuously for 27 hours. NTQO is part of the Telescopes in Education (TIE) network and therefore can be operated by students from elsewhere in the world, for example, California. In return, Australian students are able to operate TIE telescopes offshore, for example the, 24-inch at Mt. Wilson, California. Work is currently underway to use NTQO to support the educational program in Queensland schools for grade 7–9 pupils (approximately 11–15 years of age). Students undertake a project (i.e., searching for supernovae) that makes use of TIE.
Reference: http://www.eese.bee.qut.edu.au/tie/

Another Internet-controlled remote observing project is described in **Robotic scopes and research experiences for secondary students** by **Richard Gelderman**.

STARBASE is being developed to connect secondary science students and teachers with cutting-edge astronomical research. We regularly operate, via remote control over the Internet, two telescopes operating in the USA: the 0.6-m in Kentucky and the 1.3-m RCT in Arizona. Both observatories are being upgraded to provide robotic control, executing scripted observations without real-time human oversight. This telescope network is being developed in order to support our growing network of students and teachers from rural public middle and high schools in the southeast USA. Our objective is to work with the teachers to bring to their students the adventure of directly exploring the universe.

We provide professional development workshops, one at the introductory level and another in affiliation with Hands-On Universe to provide software and instruction required to introduce image-processing into the curriculum. We continue our involvement with classroom visits and the ability to request observations via our web-based interface. Most of our teachers have selected research projects involving an entire class, but we have also worked closely with individual students motivated to pursue a more detailed project (i.e., eclipsing binaries or active galactic nuclei (AGN) variability). This poster presents details of our efforts and results of the program evaluation.

A program entitled **Education project of the Japan Spaceguard Association** is described by **Yoshikawa Makoto** *et al.*

Japan Spaceguard Association (JSGA) is a non-profit organization that was founded in 1996. The principal activity of JSGA is to find and follow up Near Earth Objects by optical observations. The observations are carried out at Bisei Spaceguard Center in Okayama prefecture, Japan. When we carry out observations, we get many night sky images, which are primarily used to find moving objects. We can also use these images for educational purposes. JSGA has started to distribute them to school pupils, to let them find asteroids from these images. In order to find asteroids, we developed software called "Asteroid Catcher B-612." This software reads several images of the same area of the night sky taken some time apart, and "blinks" them so that moving objects are actually shown as moving points. We made this software easy to use. JSGA is also distributing this software with a textbook. This software and textbook are written in both Japanese and English. *Editors' Note*: Asteroid B-612 is a name from Antoine de Saint-Exupéry's delightful fable, *Le Petit Prince*.

Dill Faulkes explaining the project to students attending his former school, John Cleveland Community College. Copyright Faulkes Telescope LLC 2004.

Now we are planning a new educational project using Asteroid Catcher B-612. In this new project, pupils can analyze very recent data, and it is possible for them to find new asteroids. We call this project "Spaceguard Detective Agency."

Paul O'Brien discusses his experiences with **The Faulkes Telescope optical spectographs and Swift Satellite**.

The Faulkes Telescope project, funded primarily by the Dill Faulkes Educational Trust, has constructed two 2-m robotic telescopes located in Hawai'i and Australia. These are the largest and most powerful telescopes ever built dedicated for use by schools and colleges. We have built two optical spectrographs to be permanently mounted on these telescopes. In November 2004, NASA launched an astronomical satellite called Swift, and successful first light was reported in 2005. Swift is dedicated to the study of gamma-ray bursts, the most powerful explosive events in the universe. The Department of Physics and Astronomy at the University of Leicester has provided the X-ray camera for Swift and is a partner in the Faulkes Telescopes project. To enhance both projects, we intend to use the Faulkes Telescope optical spectrographs to study the gamma-ray bursts identified by Swift. These data will also be made available to schools, thereby raising the profile of physics and astronomy in the educational community.

(www.faulkes-telescope.com, swift.sonoma.edu)

Update at time of publishing: The spectrographs for the Faulkes Telescopes and the telescopes themselves were delayed. Faulkes North has now been commissioned and the spectrograph was installed in August 2004. Faulkes South was commissioned during spring 2005 and the spectrograph should be installed by summer 2005.

This picture of colliding galaxies NGC 2207 (right) and IC 2163 (left) was taken with the Faulkes Telescope LLC. Copyright Faulkes Telescope LLC 2004.

The project was officially launched in 2004. It has been realized through a GBP10 million donation from the Dill Faulkes Educational Trust. This trust was set up in 1998 with the aim of providing projects that would inspire young people. The Faulkes Telescope project has also received additional funding of approximately GBP750,000 from the Particle Physics and Astronomy Research Council (PPARC) for basic sponsorship, spectrographs, and staff effort; and funds from the Department for Education and Skills to enable an online educational programme to be developed. The Faulkes Telescope project is part of the National Schools Observatory, making professional telescopes available to schools.

See http://www.schoolsobservatory.org.uk/ for details. The launch event has been organized with, and funded by, PPARC.

Observational astronomy, as one of the few sciences that is truly global, is particularly conducive to international cooperation. Such issues are introduced in **Observing across continents** by **Nick Lomb** *et al*.

The dome of the Sydney Observatory Remote Telescope on the roof of the Powerhouse Museum.

Real-time use of remote telescopes can bring the excitement of professional observing into the classroom. By linking with remote telescopes across time-zones and continents it is possible to carry out observations during normal school hours. We have been collaborating to provide real-time telescope observing to students on different continents.

We have found that real-time observing is an exciting experience for the students, which gives them ownership of their observations. The presence of an observer at the telescope in text communication with the students greatly enhanced their experience.
See: http://www.powerhousemuseum.com/observe/robotic_telescope
http://www.nmt.edu/~ecorobs/

Also from Australia, **G. B. Warr** *et al.* remind us to include radio astronomy in our educational approaches in **The SEARFE radio science and astronomy awareness project**.

A pilot project to raise awareness of and gain practical experience in radio astronomy, radio science, and the radio-frequency (RF) spectrum, is being run in several Australian high schools. Students from city and country high schools are using computer-controlled RF scanners connected to wideband discone antennas (good for omnidirectional reception) to measure RF signal levels up to 1.5 GHz in their local area and are comparing their results via the Internet. Through this experience the students gain an increased understanding of the value and use of the radio spectrum for communication and astronomy, practical experience in radio science, and an appreciation of the requirements of a "radio-quiet" site for the next generation radio telescopes, the Low Frequency Array (LOFAR) and the Square Kilometre Array (SKA), which Australia is bidding to host. The effectiveness of the project is being evaluated through a number of means, including standardized tests of participating students' understanding vs. a control group, and feedback from both teachers and trained science communicators. Further details on the project can be found at the SEARFE Project website www.searfe.atnf.csiro.au.

Radio astronomy is made accessible to high-school students as described in **Contributions to radio astronomy by pre-college students** by **Michael J. Klein**.

When the Cassini spacecraft flew past Jupiter in January 2001, students and their teachers across the USA performed ground-based radio astronomy observations of Jupiter that contributed to the calibration and the interpretation of the spacecraft observations.

These students and teachers are participants in the Goldstone–Apple Valley Radio Telescope (GAVRT) science education partnership involving NASA, the Jet Propulsion Laboratory (JPL) and the Lewis Center for Education Research (LCER) in Apple Valley, California. The 34-m GAVRT radio telescope, decommissioned from NASA's Deep Space Network in 1996, is remotely controlled from classrooms across the country. GAVRT provides curriculum-based projects that include studies of the radio emissions from Jupiter and its radiation belts, observations of the microwave temperature of Uranus, support for multi-antenna Mars radar experiments, and studies of the intra-day variability of quasars. Science leadership is provided by professional scientists who establish the science objectives, interact with the students, validate data, publish the results of the GAVRT projects, and discuss the impact of student participation in the partnership.

Reflections on teaching with the aid of Internet-based robotic telescopes are included in **Teaching with Internet telescopes: some lessons learned** by **Robert Stencel**.

Observational astronomy is often difficult for pre-college students and teachers because:

(1) school occurs in daytime and visual observing occurs at night;
(2) light pollution hides the stars from students living in cities;
(3) few schools have teachers trained to use and maintain astronomy equipment;
(4) there is lack of access to expertise when needed;
(5) physically disabled students cannot easily access a telescopic eyepiece.

Internet access to computer-controlled telescopes with digital cameras can solve many of these difficulties. The web enables students and teachers to access well-maintained Internet-controllable telescopes at dark-site locations and to consult more readily with experts. This paper reports on a three-month pilot project exploring this situation, conducted Feb.–May 2002, which allowed high-school students to access a CCD-equipped, accurately pointing and tracking telescope located in New Mexico, controllable over the web, with a user-friendly sky-map browser tool. User interest proved phenomenal, and user statistics proved diverse. There were distinct lessons learned about how to enhance student participation in the research process. Details are available at www.du.edu/~rstencel/stn.htm. The authors wish to thank the ICSRC for a grant to Denver University, and acknowledge in-kind support from the estate of William Herschel Womble.

The last poster is by **Nikolay I. Perov** from Russia, who focuses on student accomplishments in the field of astronomy in **Scientifically significant astronomical discoveries from students**.

Students using a "simple measurer of angles" useful to teach mathematics and astronomy.
Photo by Rosa M. Ros.

11-year-old student modeling the apparent motion of the sun across the sky. Photo by Leonarda Fucili.

Astronomy is one of those rare subject matters that can be taught by allowing students to conduct their own research. The authors emphasize the fact that students are able to make significant scientific discoveries and develop original research about the universe, and not just reproduce well-known experiments or theories.

The authors assert that organizing teaching this way gives students a higher quality of education when compared to more traditional methods. This process – based on students generating significant works of research – has a number of aspects that are beneficial:

- problem-analytical methods are widely used;
- the students have ownership over their work;

- they experience the joy of creating their own work;
- unwanted collaboration is denied;
- their progress is tailored to their level of comfort;
- an "explosive" mastering of knowledge takes place.

Even employing research methods one-quarter of the time allows students the time to make progress towards significant scientific discoveries. The authors promote these non-traditional methods as a way to achieve highly efficient education for the pupils of Russia.

Part IV

Educating teachers

Introduction

Teachers are the key element in effective teaching and learning of astronomy. Yet very few teachers have any background in astronomy or astronomy teaching. At the elementary school level, very few teachers have any background in science at all. How much astronomy should teachers know? How should they learn it? This leads to another important issue: many teachers, especially at the elementary level, have science and mathematics "anxiety," and may transmit this anxiety to their students. It's important for teachers to have and transmit interest and enthusiasm. How can these desiderata be built into pre-service teacher education?

In Chapter 10, Mary Kay Hemenway addresses the complex topic of pre-service teacher education. Like the curriculum, teacher education varies greatly from one country to another, and even within a single country.

There are two models of teacher education: concurrent and sequential. In the concurrent model, teachers receive their content courses and pedagogy courses concurrently. The advantage is a greater integration of content and practice. In the sequential model, teachers receive a regular undergraduate degree along with hundreds of other students who are generally not prospective teachers. It may be very frustrating for prospective teachers to take science courses that are taught by the traditional lecture, textbook, and regurgitation exam method, and then to learn in teachers' college that this is not a very effective approach and that, further, this method is rarely used in schoolteaching! Of course, one of the great anomalies of the education system is that college and university instructors seldom receive any pre-service or in-service training in teaching and learning. On the other hand, another anomaly is that when the best astronomy undergraduate students graduate and want to teach school for a year or more, they are barred from teaching in public schools by their lack of education-method courses and are therefore restricted to teaching in private schools. A few programs, notably Teach for America (USA) and Teach First (UK), try to bring these top graduates into public schools in a handful of especially problematic districts, with some, but limited, success.

Especially in North America, prospective teachers may take an introductory course in astronomy for non-science students, generically called Astro 101. About 250,000 students in North America take such a course each year. Bruce Partridge and George Greenstein, in their contribution to the poster highlights for Part IV, address the issue of the goals of such a course, and how they relate to a prospective teacher's education. Partridge has been Education Officer of the American Astronomical Society. These goals were developed as a result of a series of AAS-sponsored workshops across the USA. The goals deal, as they should, with far more than the mere transfer of factual knowledge about the universe. Many new schoolteachers, however, come from teachers' colleges at which such subject-matter courses are not available.

Teachers are education professionals. As part of their professional education, they should become experts in effective teaching and learning. If they are to teach science, they should learn about effective teaching and learning of that subject. But how much content should they know? Certainly as much as the students should know! But they cannot be experts, especially in a subject such as astronomy. One of us (JRP) frequently goes into grade 6 classrooms – on condition that each student writes down one question that they would like to "ask an astronomer." The answers then go on a webpage:
http://www.erin.utoronto.ca/~astro/astrofaq.htm.

The questions rarely deal with topics that are part of the formal curriculum; they go far beyond it. There are also useful books such as Sten Odenwald's *Ask an Astronomer/Astronomy Cafe*, Neil de Grasse Tyson's *Merlin* series, and Terence Dickinson's *From the Big Bang to Planet X*.

One approach to this problem is "co-investigation." Students engage in discussion, ask high-level questions, use resources in and out of the classroom, develop research skills, theorize and explain, share and compare their theories and explanations. This approach works well with any topic, even if the teacher is not an expert. It develops lifelong, real-world learning strategies. It develops expert approaches to learning, rather than dumbing-down the curriculum. It promotes problem-based learning, and integrative, interdisciplinary learning (Woodruff 2000, http://www.oise.utoronto.ca/~ewoodruff).

The Astronomical Society of the Pacific's *Project ASTRO* is an exemplary partnership program between astronomers and teachers. The astronomers bring content knowledge and some ideas about teaching activities. The teachers bring expertise in pedagogy and class-room management. The *Project ASTRO* manual stresses the "partnership" aspects of the relationship.

Astronomers in France have been the world leaders in working with schoolteachers, as Michèle Gerbaldi describes in Chapter 11. Together, the astronomers and teachers have developed appropriate guides to content and pedagogy, activities and resources, and professional development in the form of summer schools and workshops.

One potential problem is that the material that astronomers develop may appeal to a few knowledgeable, expert teachers, but overwhelm the average (and below-average) teacher. It's important to provide material that is useful and teacher-friendly.

10

Pre-service astronomy education of teachers

Mary Kay Hemenway

The University of Texas at Austin, Austin TX 78712-1083, USA

Abstract: Although teachers are prepared in various ways to teach science, depending on the certification standards of their locality and the level at which they plan to teach, few are formally prepared to teach astronomy. In the United States, although astronomy is required for National Science Teacher certification in Earth/Space Science, and recommended for Physical Science, few teachers attempt this certification. Some certification degree programs require or recommend an astronomy course, but it is often at the introductory, non-science major level, or several weeks of astronomy within a science methods course for future elementary schoolteachers. The situation in other countries is no better. In Mexico, essentially no astronomy is taught except at the graduate school level. In South Africa, it is not taught at any teachers' college and only at some of the universities. In Portugal, it is not part of teacher preparation. In many countries, Earth–sun relations appear in the geography curriculum, but the remainder of astronomy is ignored in teacher preparation. In summary, although astronomy is found in some school curricula, teachers are often not formally prepared to teach it.

Unlike other topics in astronomy or education, there is very little research specifically on pre-service astronomy education. Perhaps it is because so few teachers are called upon to teach astronomy specifically, or because their astronomy teaching is peripheral to their main interest (e.g., general science at lower levels or physics at higher levels). Previous IAU meetings have had little information on this topic. The 1988 IAU Colloquium 105 on the Teaching of Astronomy (Pasachoff and Percy, 1990) had no papers on this topic. The 1995 ASP meeting on Astronomy Education included a panel session and one poster paper (French, 1996).

Statistics show that this problem is expected to grow. "In 1997, 1.2 billion students were enrolled in schools around the world. Of these students, 668 million were in elementary-level programs, 398 million were in secondary programs, and 88 million were in higher education programs" (NCES, 2003, Table 395). Between 1990 and 1997, except for Europe, elementary enrollment increased, for example 24 per cent in Africa. Secondary enrollment saw even greater increases: Africa (38 per cent), Oceania (68 per cent), Asia (31 per cent), Central and South America (31 per cent), North America (15 per cent), and Europe (10 per cent). Developing areas of the world had substantial increases in post-secondary enrollments: Africa (68 per cent), Asia (49 per cent), Oceania (99 per cent), Central and South America (30 per cent), Europe (15 per cent) and North America (3 per cent) (NCES, 2003). Not all the increase can be attributed to population growth; there are more opportunities for more people to obtain an education now than in the past.

With all these students attending college or university, one question might be what percentage study science and engineering. There is a wide range (13 per cent–37 per cent in

139

1999) across countries for undergraduate degrees; and an even higher range for graduate degrees, for example, from Poland (3 per cent), Italy (13 per cent), USA (14 per cent), to Japan, Switzerland, and Sweden at 42 per cent and Korea at 48 per cent (NCES, 2003, Table 412).

In spite of these increases in numbers of students attending elementary and secondary institutions, it is unlikely that many are being taught astronomy by teachers who have any formal instruction in astronomy within their own teacher preparation programs. Increases in students attending post-secondary institutions may translate into increases in students' being introduced to astronomy through a survey course, but few preparing to be teachers are required to take a specific course in astronomy, even when astronomy specifically occurs in the curriculum.

Astronomers in 14 countries (Australia, Canada, Finland, France, Germany, Mexico, New Zealand, Portugal, Romania, Slovak Republic, South Africa, Spain, UK, and USA) responded to a request for information on whether astronomy appeared in their country's schools curriculum, and whether potential teachers were required to study astronomy as part of their preparation. Only four respondents indicated that no astronomy was in the curriculum (although in some places it is taught under the title "geography"). It is not required in the teacher preparation for any country, although a few US states list it for some certification specialties. Some astronomers in the survey noted that elementary teachers often took a "science methods" course that covered the basics in an entire area of science, i.e., perhaps 45 contact hours of instruction that provide all the physical science, or life science, or earth/space science and related pedagogy that they were expected to need. Of course, they may also have taken one or more formal science courses, but these are frequently in the life sciences rather than in any area that might include astronomy. A representative from the American Institute of Physics (AIP) reported (Michael Neuschatz, 2003, private communication) that in a 1987 survey the AIP found that 3 per cent of US secondary physics teachers also taught an astronomy course, and that 11 per cent had taught it sometime in their career. The figure was so low that they removed the question from later surveys. Although astronomy is required for US National Science Teacher certification in Earth/Space Science, and recommended for Physical Science, few teachers attempt this certification. The National Council for the Accreditation of Teacher Education does not mention astronomy as a topic. A national survey in 2000 indicated that 24 per cent of grade 5–8 teachers and 34 per cent of grade 9–12 teachers had studied astronomy, as well as 44 per cent of Presidential Award winners (Horizon Research, 2001). The same survey noted that 17 per cent of secondary schools offered an astronomy course. The National Science Foundation survey, *Science and Engineering Indicators* (National Science Board 2002) showed that less than 2 per cent of US secondary students have taken an astronomy course.

Other chapters in this book examine suggested curriculum in the school setting. Generally, in elementary school, students may learn about the Earth's shape and size, its relation to the sun (day/night, seasons), and phases of the moon. The US National Science Education Standards (National Research Council, 1996) also include the solar system. Only at the secondary level do topics such as the structure of matter, or origin and evolution of the universe (including stars and galaxies), enter the US curriculum, although some states include these topics earlier in their standards.

Although "educational reform" seems to occur almost every decade, in the USA true reform of science education (including astronomy) began first with Sputnik, and later was given a boost with the publication of *A Nation at Risk* (National Commission on Excellence in Education, 1983). This document pushed teacher preparation programs to increase the fraction of time that potential teachers studied science and mathematics compared with the time spent on "educational methods." Documents such as the NSES and *Benchmarks for Science Literacy* (AAAS, 1993) followed with their vision for content and pedagogy. These standards provide expectations for all citizens, not just those planning technological careers. Changes in pedagogy have been influenced by research on how people learn (Bransford, Brown, and Cocking, 1999). Most teachers, however, were prepared for their profession under older standards. The new standards challenge those who prepare candidates as teachers with their new expectations for professional development. These standards are just now being implemented in teacher preparation programs. (If nothing else, the discussion points out the continuing need for teacher professional development programs to update current teachers in content and pedagogy.)

The UTeach program (see http://www.uteach.utexas.edu/) at the University of Texas at Austin meets these standards in secondary-level teacher preparation. Unlike previous programs in which students began their education sequence in their third year of college, and often needed a total of five years to complete a degree, this program begins with monitored experiences in schools in the first two semesters, followed by courses "Knowing and Learning" and "Classroom Interactions." "Perspectives, Research Methods, Project Based Instruction," and student teaching complete the professional sequence. The program is run jointly by the College of Education and the College of Natural Sciences. All candidates may finish in four years with a degree in science or mathematics and teaching certification. Even in this exemplary plan, astronomy is sadly lacking. Astronomy is an option that few students can fit into their tightly designed degree plan.

UTeach is but one example of thousands of avenues available to potential teachers in various countries that may prepare them to teach. Potential teachers may be trained in teacher colleges or attend general colleges and universities, as described for UTeach. In the USA, in some states they are required to obtain a degree "in a subject" with additional training in pedagogy and practice teaching for a bachelor degree. In other places, the teacher certification is an add-on year (or more) of instruction and practice teaching, sometimes leading to a master's degree.

An increasingly popular option in the USA is "alternative certification" in which a person who already has a bachelor degree, and often has many years of work experience in another career, is credentialed to teach through a series of special classes and an internship. Some programs prepare participants in special-needs areas only, such as science, mathematics, special education, or bilingual education. Studies of alternative certification have shown that, although current alternative routes may not significantly improve teacher learning, they are deemed to be no worse than many university-based teacher preparation programs (Stoddart and Floden, 1995). In the science/mathematics area, they provide teachers whose life-experiences may include actual work in science or technical fields.

One might question what makes the biggest difference in science education – where is the intersection point of students, teacher, and content? How important is content-knowledge? Ballou and Podgursky (1997) noted that many educators believe that personality, attitudes, and personal habits are just as important as professional knowledge, communication skills, and subject matter knowledge. They themselves, however, believe that teachers with stronger academic backgrounds are more effective. They found evidence that teachers from more

Fig. 10.1. Mary Kay Hemenway (center) interacting with teachers at a teachers' workshop in Sydney, July 26, 2003. Photo by Rob Hollow.

selective undergraduate institutions or who majored in the subject they teach, assigned more homework and spent longer times preparing lessons and grading. They noted a strong correlation between teachers' test scores and the scores of their students. A study by Hashweh (1997) showed that "the influence of teachers' prior subject-matter knowledge was evident in their modifications of textbook subject-matter content and through their use of explanatory representations." It was especially noted in their content organization during instruction. Those with minimal knowledge followed the textbook closely (in content and structure) and neither added nor deleted concepts. Those with a better background were likely to ask higher-order questions.

Coble and Koballa (1996) believe that "science content is the centerpiece of science teacher preparation at all levels." They point out that for most students in elementary/middle school teacher-preparation programs, this preparation consists of an average of 8.5 semester hours (2–3 courses) of the same courses taken by general education (liberal arts) students – below the US National Science Teacher's Association recommendation of 12 hours of science or science method courses for potential elementary schoolteachers. In any case, few of these courses get much beyond the survey stage and seldom lead the potential teacher to view science as a process of inquiry. Secondary teachers who major in science have more content knowledge, but seldom does their astronomy experience extend past a survey course. Trumper (2003) summarized the situation in Israel for potential elementary schoolteachers. Using a validated quiz on Earth–sun–moon relations, he found the correct response rate (36 per cent) for participants (without significant differences between first-, second-, or third-year students) did not differ for science-oriented students compared with their non-science oriented counterparts in their second or third year. (All students studied physics in their first year; 14 per cent designated as science oriented continued with more physics in their second and third years.) Upon completion of their degree, most did not have the knowledge required to teach the required curriculum.

Hanushek (1986) claimed that advanced degrees have not been found to improve teacher effectiveness, while the contribution of experience appears weak, at best, and limited to the first few years of teaching.

What does improve teachers? Induction programs – support for first-year teachers either by individual mentoring or in a support program – are becoming more common. Research has shown that science teachers participating in a group support program implemented more student-centered inquiry lessons, held beliefs aligned with student-centered practices, and felt fewer constraints in their teaching than those in other (or no) induction programs (Luft *et al.*, 2003). The US National Science Foundation noted,

Many experts assert that high-quality professional development should enhance student learning, but data for undertaking the requisite analysis are sparse. Almost all teachers participate in some form of professional development over the course of a year, most for the equivalent of a day or less. Teachers who spend more time in professional development activities are more likely to self-report improvements in classroom teaching as the result of these activities than are those who spend less time. Although several reports have asserted that teachers will perform better if they are given opportunities to sharpen their skills and keep abreast of advances in their fields, there has been no comprehensive assessment of the availability of such learning opportunities and the effects of those opportunities on teachers and students. (National Science Board, 2002, Ch. 1, p. 37).

However, NSF has been one of the principal funders of teacher professional development in areas of science and mathematics, and lifelong teacher professional development is strongly encouraged in the National Science Education Standards (National Research Council, 1996, Ch. 4) and by the National Institute of Science Education (Loucks-Horsley *et al.*, 1998).

What can astronomers do to enhance the teaching of astronomy in schools? Many have already been influential in getting astronomy into the schools curriculum. That step created a *need* for those who teach the pre-service teacher to include some, possibly minimal, astronomy in their science methods courses. A few astronomers may have connections with their colleagues in education departments and can influence those courses. But most astronomers are on one side of a great chasm that divides the science departments from the education departments. Many astronomers teach the general education course in introductory, descriptive astronomy. This is the most likely formal course that a potential teacher may take. Those who teach these courses should consider whether the course includes at least what a student should learn about astronomy by the completion of high school, a balance of conceptual understanding of Earth–sun–moon, distances and sizes, as well as the "cutting edge" topics that allow the students to understand how the tools and techniques of astronomy are used to make exciting discoveries. For those who have been prepared as teachers, astronomers can offer support. Some astronomers offer teacher professional development programs; the majority of participants in such programs have not had even a basic survey course in astronomy. (An extension of this potential aid is the involvement of secondary schoolteachers in research programs that enhance the teachers' understanding of astronomy and its methods.) The presenters of such programs should be aware of the sometimes-extensive experience the participants bring from other sciences or mathematics while holding onto naive astronomical concepts. A few astronomers write textbooks, including teacher editions, for the schools market.

In conclusion, there are no easy answers to preparing future teachers in astronomy, but it is a task worth doing. The dismal lack of understanding shown by many adults concerning basic astronomy concepts[1] is our reward for generally neglecting the preparation of those who taught them, especially in the elementary grades. Our challenge is to keep pressuring those who make policy, to forge connections with those in education departments, and to be aware of the potential impact we have on future teachers in our own survey courses.

References

AAAS (American Association for Advancement of Science) 1993, *Benchmarks for Science Literacy Project* 2061.

Ballou, Dale and Podgursky, Michael John 1997, *Teacher Pay and Teacher Quality*, Kalamazoo, Michigan: Upjohn Institute.

Bransford, John D., Brown, Ann L., and Cocking, Rodney R., eds, 1999, *How People Learn*, National Academy Press http://www.nap.edu/html/howpeople1/.

Coble, Charles R. and Koballa, Thomas R., Jr. 1996, "Science education," in John Sikula, ed., *Handbook of Research on Teacher Education*, 2nd edn, Macmillan Library Reference, 459–84.

French, L. M. 1996, "Teacher preparation," in John Percy, ed., *Astronomy Education: Current Developments, Future Coordination*, ASP Conference series, **89**, San Francisco: Astronomical Society of the Pacific, 129–30, 204–5.

Hanushek, Eric A. 1986, "The economics of schooling: production and efficiency in public schools," *Journal of Economic Literature*, **24**(3), 1141–77.

Hashweh, Marher Z. 1997, "Effects of subject-matter knowledge in the teaching of biology and physics," *Teacher and Teacher Education*, **3**(2), 109–20.

Horizon Research 2001, Report of the 2000 National Survey of Science and Mathematics Education http://2000.survey.horizon-research.com/reports/.

Loucks-Horsley, Susan, Hewson, Peter W., Love, Nancy, and Stiles, Katherine E. 1998, *Designing Professional Development for Teachers of Science and Mathematics*, Thousand Oaks, CA: Corwin Press.

Luft, Julie A., Roehrig, Gillian H., and Patterson, Nancy C. 2003, "Contrasting landscapes: a comparison of the impact of different induction programs on beginning secondary science teachers' practices, beliefs, and experiences," *Journal of Research in Science Teaching*, **40**, 77–97.

NCES (National Center for Education Statistics) 2002, *Digest of Education Statistics*, US Department of Education, Institute of Education Sciences NCES 2003–060, 2003.

National Commission on Excellence in Education 1983, *A Nation at Risk: The Imperative for Educational Reform*, US Department of Education http://www.ed.gov/pubs/NatAtRisk/.

National Research Council 1996, *National Science Education Standards*. National Academies Press http://www.nap.edu/readingroom/books/nses/html/.

National Science Board 2002, *Science and Engineering Indicators,* NSB 02–1 http://www.nsf.gov/sbe/srs/seind02/start.htm.

Pasachoff, J. and Percy, John, eds. 1990, *The Teaching of Astronomy*, Cambridge: Cambridge University Press.

Stoddart, T. and Floden, R. E. 1995, Traditional and alternative routes to teacher certification: Issues, assumptions, and misconceptions. Issue paper 95–2. National Center for Research on Teacher Learning, Michigan State University. (ERIC Reproduction Service No. ED383697) as quoted by Carol Newman and Kay Thomas http://www.aaesa.org/Pubs/99perspect/altern_teacher_certif.html.

Trumper, Ricardo, 2003, "The need for change in elementary school teacher training – a cross-college age study of future teachers' conceptions of basic astronomy concepts," *Teaching and Teacher Education*, **19**, 309–23.

---

[1] Science and Engineering Indicators 2000 National Science Foundation NSB 02–1, Ch. 7, p. 11. (http://www.nsf.gov/sbe/srs/seind02/start.htm) 86 per cent of men and 66 per cent of the women knew the Earth goes around the sun, and not vice-versa. 66 per cent of men and 42 per cent of women knew that this takes one year.

Comments

Ned Ladd: I'd like to point out a problem we encounter in getting pre-service teachers into our physics and astronomy courses. The Pennsylvania School Board Association is currently considering a recommendation that teacher candidates who have received a "C" in any undergraduate course be deleted from any applicant pool. Indeed, several local school boards have already passed resolutions to this effect. This has the unintended effect of scaring pre-service teachers from mathematics and physical science classes, where the grade distribution is generally lower. Students act rationally by choosing courses with a higher grade distribution and earn general teaching certification, even if they are interested in science teaching, rather than risking a low grade that may render them unemployable.

C. R. Chambliss: To amplify Dr. Ladd's comments, I should like to note that we are under tremendous pressure to inflate grades to maintain "standards" in our state (Pennsylvania), where education majors are now expected to receive grades of at least a "B" in all courses that they take. Given the already low standard of mathematical literacy, I am far from convinced that this is a wise idea.

Jay Pasachoff: When we use the term "astronomy" we should be explicit as to what we mean. All too often, mere discussions of Newtonian gravity are credited to "astronomy" along with phases and seasons. Perhaps we should always distinguish between "traditional" astronomy, perhaps defined as things discovered more than 300 years ago, and "modern" astronomy. We should always make room for both.

Bill Zealey: We need to ensure that in lobbying for inclusion of astronomy in upper high-school curricula we do not overstress astronomy and lose its excitement. There is a real possibility of making it just another school topic. In New South Wales, Australia, the new grade 11 and 12 high-school syllabus in physics has a large astronomy component and is in danger of turning off some students.

Mary Kay Hemenway: This is not common in most countries. Astronomy plays only a small role in the science curriculum. We should be prepared to work with the education boards of studies and not just lobby them.

11

In-service astronomy education of teachers

Michèle Gerbaldi
Université de Paris Sud XI and Institut d'Astrophysique de Paris,
98 bis Bd Arago, 75014 Paris, France

Abstract: Astronomy education of schoolteachers is reviewed in the context of in-service training when astronomy is part, or not, of the school curriculum. The methods presented are based on in-service teacher training over the past 25 years, in France. The role of a network of motivated teachers with strong links with professional astronomers is emphasized.

11.1   Introduction

Whatever the country, in general few teachers are educated in astronomy during their university studies, astronomy being an optional subject. So in-service training of schoolteachers is necessary either because astronomy is in the school curriculum or because the teachers themselves are introducing some aspect of astronomy in their lessons.

The following points will be developed:

- the context in which this training is taking place;
- the methods used for such training, taking into consideration the fact that *astronomy will be taught if the teachers feel confident.*

Examples of in-service training are taken from the French educational system because it is applied to a large body of teachers, the French curriculum being a national one, and also because in-service training in astronomy started 25 years ago through the non-profit association CLEA (Comité de Liaison Enseignants Astronomes: Teacher–Astronomers Joint Committee), created in 1977 as a consequence of the Education Commission's "Teachers Day" during the Grenoble IAU General Assembly in 1976.

In-service training has to be undertaken in two directions: one that intends to give the necessary background in astronomy–astrophysics and the other that will give to the teachers themselves the possibility of developing pedagogical resources for their needs, not forgetting that schoolteachers are also active in semi-scholarly activities: clubs, educational projects.

11.2   Overview

11.2.1   *The French school system*

In this section I give some general information on the French school system, its global organization, and the number of teachers who may teach astronomy.

The country is divided into 28 educational districts but the curriculum is national. There are 362,000 primary-schoolteachers. The primary cycle starts at age 6 and lasts five years. French secondary schools are divided into two different categories – general and technical; we shall deal hereafter only with the general one. Secondary education consists of two different successive cycles:

146

- Collège (junior high school or middle school) starts at age 11 with a duration of four years (grades 6, 5, 4 and 3);
- Lycée (senior high school or high school) starts at age 15 with a duration of three years (grades 2, 1 and Terminal [T]). At the end of the Lycée cycle, students sit for an examination to get a diploma: the Baccalauréat. Grade 1 students and those in the Terminal year can choose a scientific option for these two years of studies, so they will obtain a scientific Baccalauréat.

In the junior high schools there are: 24,800 mathematics teachers, 6,700 physics–chemistry teachers, and 11,500 biology–geology teachers; in senior high schools there are, respectively: 23,000, 13,600 and 7,600.

Not all the teachers have to teach astronomy, which is present in the curriculum only at some stages. Elementary schoolteachers seldom receive much training in science. Middle-school and high-schoolteachers receive initial training at the university in mathematics, physics, biology, or earth science, but astronomy courses are offered in only a limited number of universities.

In training colleges for the "teachers-to-be," Instituts de Formation des Maîtres (one in each educational district), some astronomy courses are given but such courses remain an exception, and the number of lectures is very limited.

11.2.2   Astronomy in the French school curriculum, in 2003

Astronomy has been present in the curriculum in primary and/or secondary education for about 20 years. In primary education, astronomy is taught through the basic topics: light and shadow, day and night, seasons, and planets. In secondary education, astronomy is taught in middle school, in grades 5 and 4, as an option that can be chosen among other topics (13 hours each year at the minimum, and 26 hours at the maximum). In grade 2 of the senior high school, astronomy is explicitly present in the programs. At this level the role of sciences in the curriculum is to inspire students to study sciences during the following two years (grade 1 and grade T) and to fight against the "science illiteracy" that is growing among that age group. For those who will in grade 1 and T choose the scientific channel, astronomy will still be present, but mainly as a consequence of the physics courses (e.g., gravitational forces, etc.).

11.3   Context for in-service training

The context in which in-service teacher training has to be organized ranges from no astronomy at all in the school curriculum to astronomy explicitly included in the school curriculum at various grades.

11.3.1   No astronomy in the school curriculum

Even in that context, astronomy may be introduced directly by the teachers themselves because astronomy is an interdisciplinary science that may be used:

- to show the pupils that knowledge is not built upon separate fields, as they may have the impression from their curriculum at school;
- to show that knowledge, in general, comes from interdisciplinary approaches.

The objective to be reached by introducing astronomy is not to put this domain of research on the front line, but to widen the student's knowledge of science, giving him or her an idea of how to work scientifically from observations and known physical laws to get new knowledge within another field of science. Nowadays there is competition with other scientific domains, such as environmental science, to fulfill these goals.

In this context, the in-service training of the schoolteachers started in France in 1977, after the creation of the CLEA, to establish a link between professional astronomers and schoolteachers. Based on that long-term experience, we can testify that, even with no astronomy in the curriculum, it is extremely useful to start the training, with a nucleus of dedicated teachers, with the following objectives:

- to create a network of trained teachers becoming *resource persons*;
- to create links among the interested teachers (newsletter, webpages);
- to exchange teaching material;
- to create partnership astronomers–teachers.

Such in-service training will be of great help, later on, to give a strong push to introduce astronomy in the school curriculum because of the fact that there is already the experience of teacher training and also because teaching astronomy has been tested by some teachers.

11.3.2   *Astronomy in the school curriculum*
Among the reasons to include astronomy in a school curriculum are:

- to offer everyone, future scientists or not, a minimum scientific knowledge in order to have a basic understanding of the world, especially as we are at a time where we have to make important decisions regarding the environment;
- to show that science differs from other fields of knowledge because it relies on scientific reasoning;
- to give every school student the opportunity to study science while trying to make science more popular among students.

With a national curriculum the number of teachers to be trained is quite large, but a new school curriculum is (or should be) introduced by the Ministry of Education with: teaching guidelines, resources for teachers (web-based, CD-ROMs) to initiate self-training and to provide educational material. These are based on previous, well-tested material that exists because the in-service teacher training started when no astronomy was present in the curriculum. There are also textbooks edited by commercial publishers specialized in education, but these were often prepared in a rush to have them ready to be put on the market when the new curriculum started. As a result, they may not be totally usable, at least for their first editions.

Even with these materials provided, in-service teacher training is requested by the teachers themselves.

11.4   Methods for in-service training
Who will do the teacher training and how? It is obvious that teacher training needs cooperation between two groups of partners: professional astronomers and schoolteachers. In any country we cannot imagine that, if a curriculum in astronomy exists, all teachers could be trained in it. It is at this point that the *resource persons* trained previously have to play an important role,

besides that of the professional astronomers, who are too few to be the only actors. Moreover, the *resource persons* can act locally in their teaching districts by personal contacts with their colleagues and informal meetings.

Various teacher training modes have been developed, each acting on different aspects of the training. Some are directed to the long-term training that aims to give full autonomy to teachers; some are much more selective concerning the subjects introduced and are efficient on a short-term basis or on only one aspect of the curriculum.

The various ways of teacher training that are mainly used are:

- summer school, about eight days;
- three one-day training sessions during the school year;
- distance learning courses;
- self-study using Internet pages, CD-ROM;
- observing sessions in an observatory.

The aims of the training are:

- to give access to theoretical knowledge through practical activities;
- to introduce the mood of observing and experimentation;
- to urge the various kinds of teachers, working at various levels, on different subjects, to exchange their experiments and to hold a dialogue; this urging is particularly important to overcome the barriers between discipline and teaching matters;
- to produce and circulate high-quality, low-cost educational material that is user friendly, not too time-consuming, and well tested from an educational point of view.

Because of the high *manpower* needed for face-to-face training sessions, these sessions cannot be provided as frequently as they should be.

11.4.1 Summer schools

It is well known that one achieves a better understanding of a subject by concentrated involvement.

A fruitful solution for in-service training is a summer school lasting eight days, bringing together teachers and astronomers. During such sessions, the teachers can also develop their critical awareness, elaborating on what can be read in books, on Internet-based resources, CD-ROM, etc.; they are able to communicate with other teachers who attend the same summer school but who teach different subjects. This point is important: the schoolteachers who attend a summer school come from various backgrounds; mixing schoolteachers with different backgrounds is enriching for everybody.

11.4.2 Three one-day training sessions

The three days of training during the school year are very often concentrated on a specific aspect of the curriculum. Such sessions can be organized in many educational districts by the *resource teachers*. Such workshops are aimed to give to the teachers the necessary basis to start a new course and, even more important, the possibility of meeting other teachers from the same district having the same concern.

It is clear, when looking back at the various training methods used, that one-day workshops are limited in their long-term impact on education of teachers. These one-day workshops, even when they are not limited to high-quality lectures, have to be considered as "refresher" courses, but they may also give, only, some *information* on a topic without leading to a lasting *transformation* at all.

11.4.3   Distance learning courses
Distance learning courses are the cornerstones for in-service education of teachers to provide a solid background.

In France, two distance learning courses (DLCs) have been created in astronomy and astrophysics through a partnership between CNED (Centre National pour l'Enseignement à Distance: National Center for Distance Learning), the Université de Paris-Sud XI, and professors of that university who are also professional astronomers (L. Bottinelli, L. Gouguenheim, and M. Gerbaldi). The first course started in 1993: Formation de Base en Astronomie-Astrophysique (Basic Training in Astronomy-Astrophysics), and the second one in 1998: Astrophysique approfondissement: à propos de l'Âge de l'Univers (Astrophysics Deepening: The Age of the Universe).

These DLCs can be granted by a university diploma. A few thousand participants have registered for these DLCs, of whom roughly 18 per cent are schoolteachers.

These DLCs are organized with printed texts, homework exercises, evaluation activities, and discussion with a tutor and the other participants using email. Two-day meetings are organized for the participants in professional observatories as well as dedicated demonstrations in planetariums.

11.4.4   Self-study using Internet pages, CD-ROM
The Internet and/or CD-ROMs do not provide full alternatives for in-service teacher training because the quality of their content is not controlled in terms of teacher training. Actually, CD-ROMs are best used as data banks or to disseminate practical exercises developed when their quality has been controlled.

11.4.5   Observing sessions
In professional observatories, small telescopes can often be used by teachers and by amateur astronomers on the basis of either a specific application or during one-week sessions. These programs give the flavor of real observations compared to other ways (using a robotic telescope, for example), but because of its cost, this latter possibility cannot be considered as widespread for in-service training.

11.5   Conclusion
Any method used in teacher education will be effective only if astronomers have developed long-term relationships with schoolteachers that nurture both teachers' astronomy knowledge and their pedagogical knowledge. The best way to achieve this joint goal is to organize summer schools from which a network of schoolteachers will emerge, trained not only in astronomical theory but also in various classroom activities.

Because these key persons are scattered all over the country, they can act locally very efficiently, in a complementary way to that of the astronomers who are grouped in a limited number of observatories or universities.

One of the major difficulties of in-service training is its cost: both for the teachers themselves (for traveling and attending summer school), and for the organization of training sessions, even for a local three-day training session for the teachers of a district, they must be free of their teaching duties at that time.

Moreover, experience has demonstrated the importance of frequent discussions as a means of transforming the basic knowledge acquired into a collaborative exchange between the participants in any training session. These exchanges involve participants from a wide range of educational settings.

But nothing will be achieved if there are no astronomers ready to provide the "gift of their time" to the goal of training schoolteachers in astronomy.

Further reading

CLEA (Comité de Liaison Enseignants Astronomes, Teacher-Astronomers Joint Committee, http://www.ac-nice.fr/clea/

Bottinelli L. 1995, "Astronomical practical activities in the new French programmes," in R. M. Ros, ed., *Fifth International Conference on Teaching Astronomy*, Barcelona: Institut de Ciències de l'Educació, Universitat Politècnica de Catalunya, 31.

Gerbaldi, M. and Xerri, A. 1996, "A multi-resource system for remote teaching in astronomy: its aims, its design, the point of view of the learners," in L. Gouguenheim, D. McNally, and J. R. Percy, eds., *New Trends in Astronomy Teaching*, IAU Colloquium 162, Cambridge: Cambridge University Press, 60.

Comments

Richard Gelderman: At Western Kentucky University, a large teacher education school in the rural south of the USA, we work with the Education School to identify pre-service teacher candidates for inclusion during professional development workshops for in-service classroom teachers. The pre-service teacher candidates provide an energetic and enthusiastic, yet non-threatening, bridge between the astronomers and the in-service workshop participants. In addition, the in-service teachers create lasting mentoring connections with the pre-service teachers.

Julieta Fierro: How do you convince astrophysicists to give the gift of their time for teacher training?

Michèle Gerbaldi: We show the colleagues the results of teacher training sessions and let the teachers themselves express their needs and explain why the participation of astrophysicists in such sessions is highly desirable.

Paul Francis: Is there any evidence that, if we teach a few keen teachers, they will spread their knowledge more widely?

Michèle Gerbaldi: In France at least, we have feedback because teachers organize training sessions in their educational districts. These sessions are advertised on dedicated webpages.

Bruce McAdam: Concerning the cooperation of astronomers in education: in New South Wales, in 1992, the Board of Studies wished to develop courses for gifted and talented (accelerated) students. Nine course outlines were developed, across all key learning areas. Only three went ahead. Eleventh year cosmology (see Chapter 3 of this volume) was successful because of the close liaison with the Astronomical Society of Australia and professional astronomers, with teachers in the writing team. The cooperation of the ASA astronomers has been crucial to the success of the cosmology unit both as guides over telescopes (full access to these) and as lecturers on real research and current ideas. They gave students tremendous access to astronomical research. In New Zealand, John Dunlop and a few teachers who were amateur astronomers prepared part of an educational curriculum K–12 to include aspects of astronomy throughout the New Zealand system. This was offered to the teams developing a new syllabus. It was professional, reasonable, and well-organized, and was largely adopted into the new K–12 courses.

James White: In the USA, the Astronomical Society of the Pacific has sponsored *Project ASTRO* for a decade. This is a program that brings professional and/or amateur astronomers together with grade 5–9 teachers. These long-term partnerships bring the astronomers into the classroom several times each year. The problem that the program normally faces is not a lack of willing astronomers but a lack of teachers.

Ruth Ernest: Responding from a comment that indicated a lack of willingness on the side of classroom teachers and preparedness to participate in training and development programs: if governments are going to keep adding to the curricula taught by teachers, they must be prepared to provide sufficient funding. In France, apparently the teachers paid for their own training (I find this completely unacceptable). Time is also another critical factor (although I have funded my attendance to the General Assembly myself, and have taken a long service leave to enable me to attend!).

Poster highlights

This section on teacher education begins with **Bruce Partridge** and **George Greenstein** asking the question **What should we teach? Goals for astronomy courses**.

Each year, more than 250,000 North American university students study astronomy. Few of these continue in the field professionally; many will go on to be secondary schoolteachers. What sorts of learning should survey courses in astronomy encourage? Two national meetings were held in 2002 to develop a list of goals for introductory survey courses in astronomy. The list of goals presented below was arrived at by consensus involving both astronomers from leading research universities and well-known science educators. While they were intended for university astronomy courses, it may be that they would be of interest also to those teaching astronomy or related physical sciences at the secondary school level. Note their generality (they were not focused on specific content items like galaxies or Newton's Laws). Nor were they intended to be a prescribed curriculum for introductory astronomy courses. Instead the set of goals developed in these meetings emphasizes deep learning, development of general skills, and good understanding of a limited number of general scientific principles, rather than broad coverage.

The full report of the meetings on goals for such courses is available in Volume 2 of the electronic journal the *Astronomy Education Review* at http://aer.noao.edu.

The goals are summarized as providing the students with:

- a cosmic perspective – a broad understanding of the nature, scope and evolution of the universe, and where the Earth and solar system fit in;
- an understanding of a limited number of crucial astronomical quantities, together with some knowledge of appropriate physical laws;
- the notion that physical laws and processes are universal – the notion that the world is knowable, and that we are coming to know it through observations, experiments, and theory (the nature of progress in science);
- exposure to the types, roles and degrees of uncertainty in science;
- an understanding of the evolution of physical systems – some knowledge of related subjects (e.g., gravity and spectra from physics) and a set of useful "tools" from related subjects such as mathematics;
- an acquaintance with the history of astronomy and the evolution of scientific ideas (science as a cultural process);
- familiarity with the night sky and how its appearance changes with time and position on Earth.

In terms of skills, values, and attitudes:

- students should be exposed to the excitement of actually doing science, as well as the evolution of scientific ideas (science as a cultural process);
- students should be introduced to how science progresses, and receive training in the roles of observations, experiments, theory, and models; analysis of evidence and hypotheses; critical thinking (including appropriate skepticism); hypothesis testing (experimental design and following the implications of a model); quantitative reasoning (and the ability to make reasonable estimates); the role of uncertainty and error in science; and how to make and use spatial/geometrical models;
- we should leave students more confident of their own critical faculties, inspired about science in general and astronomy in particular, and interested in (and better equipped to follow) scientific arguments in the media.

A statement is included here from the AAS to remind readers of the **Educational activities of the American Astronomical Society**, here represented by **Susana Deustua**.

The education activities of the American Astronomical Society are to optimize the contributions of both the AAS and its members to enhance science literacy, to encourage and broaden educational opportunities for groups under-represented in the physical sciences, and to ensure that undergraduate and graduate programs in astronomy prepare not only the next generation of research astronomers and astronomy professionals, but also broadly trained individuals with strong technical and scientific backgrounds. The education efforts include advocating greater attention to and encouragement of rewards for excellence in astronomy education as well as research on teaching and learning in astronomy, and advocating for astronomy and astronomy education in national and state education forums, to funding agencies, and to the scientific and education communities.

The American Astronomical Society Education Office maintains a website at http://www.aas.org/education that offers resources for those seeking information about AAS education programs or careers in astronomy, in addition to materials and resources on astronomy education research and for preparing education-related proposals. Two programs in which the education office is actively involved in support of the society's education goals are ComPADRE (Communities for Physics and Astronomy Digital Resources in Education) and a Workshop for New Faculty in Physics and Astronomy. Both of these are joint projects with the American Association of Physics Teachers (AAPT) and the American Physical Society (APS). The American Institute of Physics (AIP), of which the AAS is a member, is a sponsor of ComPADRE.

ComPADRE is a joint effort of the AAPT, AAS, APS, and the AIP/Society of Physics Students, funded by the National Science Foundation's National Science Digital Library program. In its first year and a half, ComPADRE has developed the infrastructure to support digital collections of materials and resources for specific communities in physics and astronomy. There are five initial collections serving different communities: undergraduate students, pre-college science teachers, informal science for the public, and resources for instructors of quantum physics and astronomy courses. Astronomycenter.org is the "Astronomy 101" collection and is a portal for teachers and students of introductory college-level astronomy to search and browse thousands of high-quality astronomy, cosmology and planetary science resources, including activities, demonstrations, tutorials, images, raw data, lecture notes, assessments, news items, and curricula. The collection contains around 2,000 items (end of 2004). We expect to increase the number of resources that have been assessed and peer-reviewed for quality and effectiveness.

Marc Gagne of West Chester University is the astronomy site editor and manages the collection. He works with an editorial board consisting of Michael Zeilik, Stephen Shawl, Laurence Marschall, Andrew Fraknoi, and Greg Bothun. More information about Com-PADRE is available at http://compadre.org.

The goal of the Workshop for New Faculty in Physics and Astronomy is to provide new faculty who are in the first few years of a tenure-track appointment with tools that will enable them to be good teachers of undergraduate students (at all levels) and graduate students. The workshop is also funded by the National Science Foundation and jointly sponsored by the American Association of Physics Teachers (AAPT), the American Physical Society (APS), and the American Astronomical Society (AAS). Participants are nominated by their department chairs. Local costs are covered by the NSF grant, but participants are responsible for their travel expenses, for example, airfare.

The Workshop for New Faculty is held each year in early November at the American Center for Physics in College Park, Maryland. The format consists of eight plenary talks given by physics and astronomy education researchers like Lillian McDermott, Eric Mazur, and Tim Slater, as well as a dozen breakout sessions. Topics covered include: Research as a guide to improving student learning, Active learning and interactive lectures, How to help your students develop expertise in problem solving, Are your students learning? How do you know?, How to get your students to prepare for every class. The breakout sessions expand on the talks, and aim to provide some practical exposure to such teaching techniques as using peer instruction and lecture-tutorials in large lecture classes. This arrangement seemed to work well, as evidenced from the mostly positive survey responses, and the waiting lists. In 2003, there were about 120 participants, of which 50 per cent expect to teach introductory astronomy courses, although only ten identified themselves as AAS members.

The difficult job of integrating **Astronomy education in a primary teacher training institute** is addressed by **Bill MacIntyre**.

Teacher self-efficacy for pre-service teachers in astronomy education is dependent on subject-matter knowledge and pedagogical content knowledge (PCK), as well as an understanding of educational practices and curricular knowledge. Astronomy education at Massey University College of Education (New Zealand) is offered at three different levels. The basic level is 2–4 hours of training in a 40-hour compulsory science methods course. The intermediate level is 6–10 hours of training in a 40-hour optional science methods course while additional astronomy education is offered in a 40-hour "optional" studies in subject course "Spaceship Earth and Beyond."

A study by Lewthwaite and MacIntyre (2003) asked the question, "Does the strength of efficacy in teaching a phenomenon (understanding the pedagogical content knowledge) correlate to conceptual understanding of the phenomenon?" Pre-service students who took the course "Spaceship Earth and Beyond" in addition to the compulsory science methods paper showed a statistically positive correlation between their belief that they could teach astronomical concepts effectively and their own astronomy understanding. The positive correlation is due mainly to the fact that "Spaceship Earth and Beyond" specifically addresses the subject matter knowledge necessary to teach the astronomy in the primary school science curriculum and the pedagogical content knowledge using an "investigating with models" approach.

The development of astronomy understanding is achieved through the collection of evidence (data) from night sky measurements, personal observations, and computer simulation (*Starry Night*), which students then model using 3D models to support or refute their explanations of astronomical events.

Further reading

Lewthwaite, B. and MacIntyre, B. 2003, "Professional science knowledge and self-efficacy: a vignette study," in *SAME papers 2003*, ed. R. K. Coll, Hamilton, NZ: Centre for Science and Technology Education Research, 161–88.

Astronomers and astronomical organizations often produce resource material for schoolteachers, but is the material usable and effective? **John Percy, Cresencia Fong, and Earl Woodruff (Canada)** ask: **What do teachers see in an "exemplary" astronomy video?**

Their study explored the use of video presentation of exemplary astronomy teaching, as a teacher professional development tool. A qualitative/quantitative hybrid design was used to collect data on the video elements that 11 pre-service and 11 experienced teachers attended to as they observed such videos. The elements were classified according to whether they reflected the subject content, the management of the demonstration in the classroom, the production elements in the video, or the true pedagogy or exemplary practice. The results suggested that teachers rarely recognized the exemplary practice presented to them, especially if they were inexperienced teachers, and/or if they were unfamiliar with the astronomical content. There was some evidence to suggest that instructing teachers to seek out exemplary practice aids them in attending to the underlying functions of the methodology presented. The study provides one more example of a general problem in science education and communication: what learners absorb may not be what the educator thinks or hopes that they will absorb!

The challenge of educating teachers is further addressed in **A space science teacher professional development program** by **Sanjay S. Limaye**.

Recent adoption of state/national science education standards by school districts in the USA has created a need for effective teacher professional development in space science at elementary-, middle-, and high-school level. Particularly at the elementary- and middle-school levels, the majority of teachers teaching astronomy/space science content have had little education in the area, regardless of when they obtained their certification. To meet this growing need, the Office of Space Science Education has developed a program to offer teachers background content knowledge through summer workshops and periodic school year meetings for a small number of teachers from Wisconsin and Illinois.

The program has included lectures by experts, tours of observatories (professional and amateur), science museums, planetariums, and online learning. A highlight of the program has been introducing teachers to hands-on observing through remotely accessible telescopes.

Another aspect has been to make them aware of the many resources available to them through NASA missions. The most significant benefit for the teachers, however, has been the creation of a peer group and the support it offers in sharing curriculum and lesson plans. This effort has been supported by a NASA/IDEAS grant.

An interdisciplinary approach to tackling subjects in astronomy is introduced by **Janelle M. Bailey** in a paper entitled **An innovative astrobiology K–12 Teacher Enhancement Program**.

Demonstrating pre-service "teacher trainees" modeling the "evidence" to support one notion as a more appropriate explanation for the cause of seasons.

Demonstrating the position (high in sky) of the sun in summer and the shadow formed by an object.

Demonstrating the position (low in sky) of the sun in winter and the shadow formed by an object.

Demonstrating the shadow formed at both equinoxes.

Teachers at the 2002 Summer School of the European Association for Astronomy Education in Finland, doing "daytime astronomy." Photo by Rosa M. Ros.

Pre-service teachers busy making simple hands-on materials for teaching astronomy at the elementary-school level. Photo by Leonarda Fucili.

The 2003 Towards Other Planetary Systems (TOPS) Teacher Enhancement Program represents the final summer offering of a five-year US National Science Foundation-funded project at the University of Hawai'i Institute for Astronomy. Focused on high-school science teachers from the Hawai'ian and Pacific Islands, the workshop is intended to use the theme of cutting-edge astronomical research to motivate and train high-schoolteachers to enhance the quantity and quality of astronomy instruction in the classrooms. Teacher-participants attend three weeks of lectures by professional astronomers, tour world-renowned facilities, work with science educators using nationally recognized hands-on astronomy curricula (primarily but not exclusively from *Hands On Astrophysics*), are introduced to astronomy data and teaching resources, study alternative student assessment strategies, participate in group discussions about integrating astronomy into their classrooms, complete telescope observing projects, and, during the last two weeks of the workshop, thanks to a private donor, work side by side with high-school students from the Hawai'ian and Pacific Islands.

Another submission from **Janelle M. Bailey** recounts one of the more successful programs of integrating astronomy into the US college curriculum in **Teaching introductory astronomy using lecture-tutorials**.

Bailey and her colleagues have developed a series of innovative classroom instructional materials for introductory astronomy courses. The materials package, called *Lecture-Tutorials for Introductory Astronomy*, is a self-contained, classroom-ready product for use with collaborative student learning groups. The materials are designed to be easily integrated into the conventional lecture course and provide faculty with effective, student-centered, classroom-ready materials that do not require a drastic course revision for implementation.

Topics include, for example, the nature of light and the electromagnetic spectrum; motions in the solar system; techniques in astronomy; stellar evolution; and cosmology and the big bang. In "Looking at distant objects," students first reason about signals sent from nearby stars and the time involved in such a transmission; later they turn their attention to more distant objects and determine the chronology for the observation from Earth of a supernova explosion in a distant galaxy. Each activity takes approximately 15 minutes, and students are asked to reason about difficult concepts in astronomy while working in pairs. The materials are based upon research on student misconceptions, effective instructional strategies, and extensive field-testing.

Federal funding was provided by NSF CCLI #9952232 and NSF Geosciences Education #9907755. *Lecture-Tutorials for Introductory Astronomy* is published by Prentice Hall, United States, ISBN 0-13-101109-X.

It is often difficult for teachers to communicate with regards to teaching methods and curriculum, especially in underdeveloped countries. A perspective from Thailand is provided by **Busaba Hutawarakorn** *et al.*

They report the latest development of a pilot project in establishing an astronomical network for teachers in Thailand. The project is funded by the Institute for the Promotion of Teaching Science and Technology, Thailand, and operated by Sirindhorn Observatory at Chiangmai University. The objectives of the project are to establish semi-robotic telescopes that can be accessed from schools nationwide, and to establish an educational website in the Thai language (www.astroschool.in.th) that contains education resources and links to other educational websites worldwide. The network will play an important role in the development of teaching and learning astronomy in Thailand.

Part V

Astronomy and pseudoscience

Introduction

The prefix "pseudo" is derived from a Greek word meaning "false." Pseudoscience refers to theories, assumptions, and methods that are mistakenly thought to be scientific. Obviously the distinction between science and pseudoscience is not clear-cut. Pseudoscience overlaps with scientific "misconceptions" – beliefs that are held by students and the general public that differ from scientific fact. Both of these concepts are subtle ones, since all scientific theories can be regarded as tentative to some degree.

At the 1996 IAU conference on astronomy education in London, UK, Neil Comins classified astronomical misconceptions into about 20 types, and he has described these in detail in his book *Heavenly Errors* (Columbia University Press, 2001). Some misconceptions are cognitive in nature; they are dealt with in our Part II. Others are products of religious belief or superstition (often transmitted through the "authority" of family or friends), or popular culture, or errors or excesses of the media. In relation to astronomy, notable pseudosciences include astrology, space aliens, and creationism; the works of Velikovsky also fall under the pseudoscience rubric. The majority of students will be affected in some way by these beliefs.

How to deal with them? The theory of constructivism is one of the most influential science-learning theories in schools today. It states that, by reflecting on their own knowledge and experiences, students build their own new knowledge about the universe around them. Teachers must therefore be aware of students' pseudoscientific beliefs, as well as their other misconceptions, if they are to correct them through their teaching. Many of them are deeply rooted. They cannot easily be changed by lectures and textbooks. They must be confronted through minds-on teaching.

To Jayant Narlikar's eloquent chapter we have added a short section on pseudoscience in North America – partly because it is rampant there, and partly because astronomers and others have developed resource material to help teachers deal with this issue. These include the magazine *Skeptical Inquirer*, published under the auspices of the Committee for the Scientific Investigation of Claims of the Paranormal (part of the international "skeptics network"), and the excellent webpages maintained by the Astronomical Society of the Pacific through the efforts of Andrew Fraknoi (http://www.astrosociety.org/education/resources/pseudobib.html).

It was suggested at the conference that the IAU Commission on Education and Development take a more active role in fighting pseudoscience. We are open to suggestions from readers as to how this can best be done.

12

Astronomy, pseudoscience and rational thinking

Jayant V. Narlikar

Inter-University Center for Astronomy and Astrophysics,
Post Bag 4, Pune, 411007, India

Abstract: A strong case is made for including astronomy in the school science curriculum, as it encourages a scientific outlook. The realization that awesome natural phenomena can be explained in terms of known science can develop in students the habit of thinking rationally and help them counter superstitions that have traditionally taken root in society. A contrast with a pseudoscience like astrology will further help them to come to grips with the way real science functions.

12.1   Introduction

In 1944, three years before India became independent of British rule, Jawaharlal Nehru wrote in his now famous book *Discovery of India*:

The impact of science and the modern world have brought a greater appreciation of facts, a more critical faculty, a weighing of evidence, a refusal to accept tradition merely because it is tradition ... But even today it is strange how we suddenly become overwhelmed by tradition, and the critical faculties of even intelligent men cease to function.

He then went on to express the hope that "Only when we are politically and economically free will the mind function normally and critically." India became independent in 1947 with Nehru as the first Prime Minister, a post that he held for nearly 17 years. Ever an advocate of science and technology as the means of progress, he encouraged establishment of a good scientific infrastructure and also looked after achieving industrial growth. However, what has been the net outcome so far as human resources are concerned? Now we are well into the sixth decade after independence: where do we stand vis-à-vis Nehru's expectations of rational thinking?

12.2   The present scenario

The picture is not very encouraging! Here are some pointers to how the social mind-set operates today, both collectively and individually:

(1)   Many marriages, even among educated graduates, are decided after the matching of horoscopes. In India, arranged marriages are still the norm, although the prospective bride and groom have greater say in the process. All the same, one condition that many families insist upon is that their horoscopes should be compatible! Many matches, otherwise suitable, are not allowed to take place if the planets disapprove.

(2) Three years ago, the state of Uttar Pradesh was further subdivided to create another smaller state called Uttarakhand. The date and time of the ceremony for launching the new state was fixed after consulting astrologers. The ceremony was in fact postponed by a few days until an "auspicious period" approved by astrologers began. Likewise, it is not uncommon for state or union cabinets to be sworn in during auspicious periods.

(3) In 2000, when several planets happened to be in the same part of the sky (i.e., close to, but not exactly in alignment), doomsday was predicted. It was claimed that all sorts of catastrophes, including volcanic eruptions, tidal waves, etc. would occur. This prediction led an entire seaside village in the state of Gujarat to go into a mind-set of panic, with people abandoning their homes and running away. Nothing untoward happened, of course, except that burglars made easy pickings from the abandoned houses.

(4) During the last decade a new pseudoscience under the name of "*vastu-shastra*," literally translated as "science of architecture," is taking hold of the urban middle and upper classes. This subject lays down a series of rules for designing and constructing houses in relation to the ambience. While architects do have a set of rules that seek to relate the environmental factors to the siting of various rooms (e.g., in cold countries of the northern hemisphere, the south face is expected to capture the maximum sunlight and accordingly rooms with south-facing windows are planned), the dictates of *vastu-shastra* do not have any logical connection to environment. Here are some instances of what these rules recommend for the housing site:

> If the boundary lines of the plot on which the house is built are not parallel to the magnetic axis, such land is poor for overall growth, peace and happiness.
> If the land is high in the north or east directions, it causes financial losses or can damage the prospects of the owner's male children.
> Avoid sites shaped like a triangle as they will lead to government harassment, while a parallelogram can lead to quarrels in the family.

> And so on and so forth! These suggestions are taken seriously, and the clients often make the architect change the plans, even demolishing what has already been built. *Editors' Note*: The Asian practice of *feng shui*, apparently similar in nature, is unfortunately becoming popular in some circles in the USA.

(5) In 1995, there was a miracle! The idols of Ganesha in several temples all over India began to drink milk if fed through their trunk. (Ganesha is the elephant god, who has the head of an elephant on a human body.) Believing the miracle, huge crowds gathered in these temples. The phenomenon was explained by scientists using the concepts of surface tension and capillary action. Yet many educated persons to date believe that the idol was permeated by a divine presence who drank the milk.

(6) In the year 2001, the apex body of higher education in India, the University Grants Commission, declared astrology a science and offered funds for instituting its teaching in universities! After protests by scientists, the UGC withdrew the claim that astrology is a science and instead placed the subject in the humanities stream. It was claimed by the UGC circular that astrology provides understanding of time, and that it is useful in business as well as several other fields.

These examples suffice to demonstrate that the goal of rational thinking is still some way away from the mental make-up of the average Indian. The question is whether something can be done to address this dismal situation.

12.3   The teaching of astronomy as an antidote

The disturbing fact about present symptoms is that the trend towards superstitions and beliefs in pseudosciences is growing and the younger generation is getting more and more attracted to pseudosciences. One way to start rectifying the situation is to demonstrate at the school level the way real science works and to contrast it with pseudoscience. Take the following example of chemistry:

In India, non-governmental organizations devoted to eradication of superstitions have two visiting teams doing the rounds of various schools. The first team performs "miracles" like producing sudden changes of colour, fire eating, etc., which the second team, following it, explains by the demonstrations of on-site chemistry experiments. When the unexpected or the unusual gets demonstrated as a known fact of science, the belief in miracles crumbles. Such direct demonstrations have proved to be very useful and effective. The Ganesha phenomenon was likewise demonstrated and clarified by scientific experiments. Although some die-hard believers were unmoved, schoolchildren at least found the debunking effective. I believe education in astronomy right from the school level can play a similar role in this enterprise. For astronomy provides similar instances that distinguish it as a science from pseudoscience. These instances come through several channels, which can be incorporated into school science curricula. Before discussing some examples, I wish to mention that normally astronomy appears in school texts as an extra chapter in geography! The planetary system, the Milky Way galaxy, and the universe are tucked in as the last lesson after pupils have taken in the different geographical aspects of the countries and regions on the Earth. Thus the lesson hardly presents astronomy as a science.

(1) **Solar and lunar eclipses**. These are relatively rare phenomena but are spectacular (especially solar eclipses) when they do take place. An eclipse may not be visible directly at a given location, but thanks to information technology it can still be seen, on a computer screen, as occurring at another part of the globe. Traditionally, in India eclipses have been considered sinister, with demons swallowing the sun and the moon. At the solar eclipses of 1980 and 1995, for example, cities like Bombay wore a deserted look, since it was considered dangerous to venture out when the sun is being swallowed! Today, a secondary-school student can appreciate that eclipses are mere shadow play. The students can see how it is possible to calculate and predict when an eclipse will occur. One should explain the precautions to be observed while viewing a partially covered sun. One can also take this opportunity to debunk some long-standing superstitions, for example, the belief that it is dangerous for an expectant mother to be out under the eclipsed sun.

(2) **Transits and occultations**. School students have a passing knowledge of Newton's law of gravitation. They can appreciate how the law not only governs the fall of an apple but also the motions of planets and satellites. A transit of Venus across the solar disc (June 8, 2004 and June 5/6, 2012) can therefore be exactly timed, which can be demonstrated as a scientific experiment.

(3) **Tides**. Students living in seaside places are familiar with tides. They can be made to time high and low tides and relate them to the position and phase of the moon. Again the law of gravitation can be brought up to explain the phenomenon.

Indeed the message that can be clearly communicated is that astronomy is a branch of science that epitomizes humanity's attempt to understand the universe in terms of the

Fig. 12.1. A statue of Galileo Galilei (1564–1642) at Inter-University Centre for Astronomy and Astrophysics in Pune, India. Photo by Jay M. Pasachoff.

science studied here on the Earth. At a more sophisticated level, one can bring up discussions of why the sun shines, why different stars have different colors, how astronomers can study the universe with non-visual forms of light, and so on. But the bottom line should be that one must do one's best to understand natural phenomena with the tools of science.

12.4   A critical look at astrology

In the Indian context, special efforts are needed to counteract the growing influence of astrology. To contrast a pseudoscience like astrology with hard science like astronomy, the above examples from astronomy may be put side by side with how astrology operates. Thus children could be asked to apply the following criteria to astrology:

- Scientific predictions are falsifiable and are therefore worded in a precise manner. Are astrological predictions so worded or are they vague and could be made consistent with any result? As an example, children may be asked to compare the forecasts for different zodiacal signs that appear in different newspapers and see if there is any consistency in them.
- Even if the astrological forecasts are tested for success or failure, how often are they successful and how often not? The example of coin-tossing may illustrate the circumstance that a prediction may turn out correct purely by chance. So if you correctly predict 50 times out of 100, are you able to claim any special knowledge or special predictive power? At the high-school level, some account of probability and statistics may be brought in to quantify the issues.
- Is the making of a prediction logical and based on some precisely stated assumptions? All too often astrologers are *ad hoc* in their approach, and change the premise *post facto* if the forecast does not tally with facts. It can be explained that science does not permit this luxury. Ideas of Karl Popper regarding falsifying a scientific theory can be explained.
- Do the successes/failures of a prediction depend on who makes them? That is, is there objectivity about them? When a particular astrological prediction fails, other astrologers point a finger at the person who made the prediction, stating that he is not competent. In short, one encounters here a subjective element which is not permitted in science. If a scientific experiment cannot be repeated by other labs, it loses credibility.
- Field tests on large samples and their overall success-evaluation have established the validity of empirical relationships in physics, which were subsequently explained by basic laws. What has been astrology's record with such field tests? It turns out that astrologers discourage such field tests. However, such tests as have been conducted by social scientists and statisticians show that the success rate of an astrological forecast is statistically insignificant.

Ample examples exist in literature in support of the above points and to show why astrology is not a science. It will be worthwhile spending some part of astronomy education in demonstrating these facts to children.

12.5   Concluding remarks

Science teaches rational thinking, and the application of the scientific outlook goes well beyond the laboratory walls. Astronomy boldly applies the laws of science to the grand laboratory that is our universe. Some natural phenomena can be awe-inspiring and, if not explained scientifically, can lead to superstitions and irrational thinking. Children could learn how to think rationally and not get carried away by superstitions, pseudoscience, and loose-thinking, if they are given some introduction to astronomy as a science and are encouraged to apply scientific criteria to see why astrology is not a science. Indeed, at the school level, it may be possible to encourage children to argue with their parents and try to debunk their beliefs! The school may set up a parent–teacher forum wherein such ideas can be freely discussed. It may be worthwhile involving parents also in the viewing of astronomical phenomena like eclipses, transits, or even routine night-sky watching. In short, there is considerable scope for the introduction of astronomy in the school curriculum, but in the science stream and not as an appendage to geography!

Comments

Martin George: Given the far stronger following of astrology in India, does this make it harder to teach the difference between astronomy and astrology in schools?

Jayant Narlikar: This makes it all the more important to teach it! But in India, because of the stronger following, it is necessary to get the message through to the parents of schoolchildren too. So, yes, this makes it twice as difficult.

Case Rijsdijk: Astrology is on the rise not only in India but around the world – do you have an explanation for this?

Jayant Narlikar: Life-style as a whole is getting more competitive and stressful. There are often situations when a person is at a loss to respond properly. At such a stage there is temptation to leave the decision-making to someone else. Astrology or astrologers claim to provide answers. Even though they are not based on rationality or facts, the so-called answers attract the person especially under a stressful condition. This may provide a possible answer to the question.

John Hearnshaw: You noted that superstition in India has increased since the end of British rule in 1947. Does this mean you advocate a return to British rule?

Jayant Narlikar: No! Of course not! It has been found in other instances that in a repressive regime, free thinking is suppressed, and when the repression is lifted, all different thoughts and beliefs, including superstitions, spread more freely. In the Indian context, one major factor contributing to growth of superstitions is the information technology revolution that has spread pseudoscience rapidly.

Allan Kreuiter: There is a major flaw in Western astrology that precession has changed 12 zodiac signs into 13. Are there similar Indian equivalents, and what is the response to pointing out astronomical flaws in astrology?

Jayant Narlikar: Indian astrology has a Greek, not vedic, tradition that is not known by the public and not appreciated when explained.

John Baruch: One serious pseudoscience we have to deal with in the UK to an increasing degree is creationism. Do you have this problem in India? If so, how are you dealing with it?

Jayant Narlikar: This is only one of many problems we face. There is a website at the AAS put together by a group of four that deals with this (see Chapter 13 of this volume).

T. K. Menon: It is not only in India that beliefs of government officials can influence actions affecting service. For example, in a recent proposal for an accelerator in a Canadian province, the reporters were asked to remove all references to "Age of the Universe" because the Head of Government felt that he knew the answer to be 4,000 years.

Fig. 12.2. Statue of Aryabhatta (476–550), an Indian astronomer/mathematician, at Inter-University Centre for Astronomy and Astrophysics in Pune, India. Photo by Jay M. Pasachoff.

Editor's Note: Indian astronomer Aryabhatta is thought to have computed pi to five decimal places, to have computed a table of sines, and to have proposed, as Jayaram Chengular of the Giant Metrewave Radio Telescope described, "(1) a heliocentric theory, (2) that the earth spins on its own axis, (3) that moonlight is reflected sunlight, and (4) that eclipses are caused by the shadow of the moon, and to predict eclipses."

Aryabhatta is also the author of the first of the later siddhantas called *Aryabhattiyam*, which sketches his mathematical, planetary, and cosmic theories. This book is divided into four chapters: (i) the astronomical constants and the sine table, (ii) mathematics required for computations, (iii) division of time and rules for computing the longitudes of planets using eccentrics and epicycles, (iv) the armillary sphere, rules relating to problems of trigonometry and the computation of eclipses.

The parameters of *Aryabhattiyam* have, as their origin, the commencement of Kaliyuga on Friday, February 18, 3102 BCE. He wrote another book where the epoch is a bit different.

Aryabhatta took the Earth to spin on its axis; this idea appears to have been his innovation. He also considered the heavenly motions to go through a cycle of 4.32 billion years; here he went with an older tradition, but he introduced a new scheme of subdivisions within this great cycle.

That Aryabhatta was aware of the relativity of motion is clear from this passage in his book, "Just as a man in a boat sees the trees on the bank move in the opposite direction, so an observer on the equator sees the stationary stars as moving precisely toward the west."

The first Indian experimental space satellite, launched in 1975, was named after him.

Reference

www.hindunet.org/science_after_aryabhatta/

13

Astronomical pseudosciences in North America

John R. Percy and Jay M. Pasachoff

(Percy) University of Toronto, Toronto, ON, Canada
(Pasachoff) Williams College, Williamstown, MA, USA

Abstract: We briefly comment on three astronomical pseudosciences – astrology, creationism, and space aliens – which are accepted in North America by a significant fraction of the population. We list articles, books, websites, and organizations that provide useful resources which teachers, students, and the public can use to deal with such beliefs.

13.1   Pseudoscience

Pseudosciences are widely accepted in India, as Jayant Narlikar has explained in the previous chapter. Pseudosciences are also widespread in other populous countries that are not part of the "first world." But pseudosciences also became very popular in countries such as Russia after the breakup of the Soviet Union. Science education was generally considered strong in these countries. Perhaps the coming of "democracy" encourages people to believe what they want (or for their beliefs to become more open). In North America, several forms of astronomical pseudosciences are widely accepted. Fortunately, there are individuals such as Andrew Fraknoi, and organizations such as the Astronomical Society of the Pacific (with which Fraknoi is associated), which have produced some excellent resource material for teachers and the general public; these are online at:

http://www.astrosociety.org/education/resources/pseudobib.html

The Committee for the Scientific Investigation of Claims of the Paranormal (CSICOP), based in Buffalo, New York, and with a website at http://www.csicop.org, describes itself as follows: "CSICOP encourages the critical investigation of paranormal and fringe-science claims from a responsible, scientific point of view and disseminates factual information about the results of such inquiries to the scientific community and the public." They publish a valuable magazine, *Skeptical Inquirer*, six times a year. An index of past articles is on line at their website.

Confronting pseudoscience brings to mind many books by Martin Gardner, ranging back to his *Fads and Fallacies in the Name of Science* (1957). More recent works on related subjects includes books by Gardner (such as *Did Adam and Eve Have Navels?* and *Discourses on Reflexology, Numerology, Urine Therapy, and Other Dubious Subjects*, W. W. Norton, 2000); James Randi's *Flim-Flam! Psychics, ESP, Unicorns, and Other Delusions*; Michael Shermer's *Why People Believe Weird Things*; and Ann Druyan and Carl Sagan's *Demon-Haunted World*.

13.2   Astrology

Astrology is the belief that human traits and behavior are somehow connected with the positions of the sun, moon, and planets in relation to the stars. Astronomy and astrology were associated for two millennia, but parted company in the Renaissance. Nowadays, astronomy and astrology are frequently confused; most astronomers have received mail addressed to the "Department of Astrology"! It is the pseudoscience that is most widely associated with astronomy.

At the conference on which this book is based, Case Rijsdijk (South Africa) suggested that, for many people, astrology is a belief system. Because it has significant cultural connotations, it needs to be treated far more carefully and differently, compared with issues such as green men from Mars, UFOs, perpetual motion machines, the Bermuda Triangle, Planet X, and assorted doomsday stars.

De Robertis and Delaney (1993, 2000) have published two surveys of the attitudes of students at a large Canadian urban university to astrology and astronomy. In the second survey, they find that over half of both arts and science students subscribe at least somewhat to the principles of astrology. Also, 60 per cent of arts students and 50 per cent of science students are unable to distinguish between astronomy as the science and astrology as the pseudoscience. Females are much more likely to subscribe strongly to pseudoscience than males from the same faculty, and with equivalent mathematics background. In the first survey, the responses of the science students tended to be more skeptical; in the second survey, there was no significant difference. In both of their articles, the authors make some suggestions about how the education system can deal with this issue.

Earlier, one of us (JMP) along with Richard J. Cohen and Nancy W. Pasachoff (Pasachoff *et al.*, 1970) surveyed Harvard undergraduates for comparison with students from West Africa in relation to belief in the supernatural. Pseudoscientific beliefs, high then, are not likely to be lower now. What this article discusses that is particularly worrisome is that pseudoscientific beliefs can lead to harm to the general public. For example, there were old reports of people refusing vaccination; most unfortunately, there are current (2005) outbreaks of polio in Nigeria because of a pseudoscientific belief. Opposing pseudosciences in general may, therefore, have significance beyond mere intellectual reasoning.

The usual prescription for dealing with pseudoscience is to teach students the general principles of rational thinking and the scientific method. In the case of astrology, however, it may be difficult for students to appreciate the evidence. They may read about supporters of astrology in the newspapers; they will certainly see horoscopes there. (Newspaper editors justify publishing horoscopes by saying that they are for entertainment only.)

Further, recent education research points to the need to overcome student misconceptions in a wide variety of fields. There has been much discussion (by Philip M. Sadler and others) of how merely teaching the correct material doesn't reach the heart of student understanding. Students' misconceptions are deeply rooted. They may arise from faith in the teachings of their religion, from the authority of their parents or other elders, from the myths and superstitions of their culture, or from so-called "common-sense." Neil Comins, in particular, has written a book about misconceptions (Comins, 2001, 2003) and given many examples. One of us (JMP) is not alone in adopting the habit of listing misconceptions on given topics in his textbook and accompanying Web material (http://info.brookscole.com/pasachoff). In any case, it is not a surprise that teaching valid astronomy doesn't debunk astrology in people's minds.

Many different studies of the validity of astrology have been published. Some of them survey lists of eminent individuals (artists, for instance) to see whether they are more likely to be born under a birth sign associated with a relevant trait – creativity, for instance, in the case of artists. No significant correlations exist. Carlson (1985) published the results of a double-blind study of the efficacy of astrological predictions. Astrologers assisted in validating the study. Again, there was no significant positive result.

One interesting classroom approach is that of Wynn-Williams (1998). He gives students copies of the 12 horoscopes from the previous day's newspaper. The sun signs are removed, and replaced by code numbers that are not known to the students, and that make the analysis of the results quick and easy. Students are asked to choose the one of the 12 horoscopes that would have been most suitable for yesterday. Predictably, the results do not differ from chance. The students are introduced to "the statistics of small numbers." Astrologers would complain that sun sign horoscopes are unduly simplistic. But the students are exposed to the scientific method of experimentation and analysis.

People can benefit, of course, from reading one or more horoscopes, if they contain good advice. Likewise, advice from astrologers can be useful, or at least comforting, if they are good social workers. But, "it's not in the stars."

13.3   Creationism

According to the results of a 2001 Gallup poll, 45 per cent of Americans believe "God created human beings pretty much in their present form at one time within the last 10,000 years or so" and, in a forced choice between "the theory of creationism" and the "theory of evolution," 57 per cent chose creationism against only 33 per cent for evolution (quoted by Shermer, 2002). Students undoubtedly acquire these beliefs from their religion and culture, and especially from their family and friends.

There is a large and well-funded movement in the USA to discredit evolution, and to lobby for "equal time" in the school curriculum, and in textbooks, for evolution and for creationism or for "intelligent design" – presumably as a first step to eliminating evolution completely. Fortunately, scientists and educators have strongly opposed such moves. They have created resources that concerned citizens can use in any part of the country where creationism rears its head. In this way, those concerned citizens need not "reinvent the wheel." In the biological sciences, and to some extent in the earth sciences, scientists and educators have produced materials that teachers can use in their classrooms to deal with this issue. The National Center for Science Education (http://www.ncseweb.org) plays an important role. Until recently, however, there were few or no resources for teachers to use to explain how astronomers know that the universe is very old, for example, and evolving in time.

In 2000, concerned by the tendency to de-emphasize the teaching of evolution, the President and Council of the American Astronomical Society (AAS) issued a formal statement on behalf of the astronomical community. Then, on behalf of the AAS Education Board, Andrew Fraknoi, George Greenstein, Bruce Partridge, and John Percy created "An ancient universe: how astronomers know the vast scale of Cosmic time." This document is aimed at teachers at the grades 6–12 level, i.e., middle school and high school. It contains information, lists of resources, and links to exemplary activities. It was published as a special edition of the Astronomical Society of the Pacific's *The Universe in the Classroom*, a quarterly newsletter for teachers. It is available at: http://www.astrosociety.org/education/publications/tnl/56/.

Wesley Wildman, associate professor of theology and ethics at Boston University, lectures about "Rationality in science and religion: a pragmatist approach." He points out that the secularist theory of the 1960s, which predicted that religion would diminish by the end of the century, is not borne out by data, certainly not consistently around the world. He worries that too many of the scientists' retorts to creationists take a condescending tone that is not successful in accomplishing its aims. Wildman and Mark Richardson, a professor at the General Theological Seminary in New York, edited *Religion and Science: History, Method, Dialogue*, a compilation of essays that discuss the relationship and conflict between religion and science.

13.4   Space aliens

According to a 2001 National Science Foundation survey, 30 per cent of respondents agreed that "some of the unidentified flying objects that have been reported are really space vehicles from other civilizations." About the same percentage of a Gallup poll reported that they believed that "extraterrestrial beings have visited the earth at some time in the past." Students probably acquire these beliefs from the media; family and friends may also have some influence.

In addition to the usual prescriptions for dealing with pseudoscience, one of us (JRP) has found that students take note when told that, according to a survey, about 2 per cent of Americans believe that they have been kidnapped by space aliens. That translates into the abduction of about 6,000,000 Americans – without anyone noticing. This is sometimes referred to as the "Santa Claus problem," a reference to the difficulty that Santa Claus would have in visiting every house on Earth on Christmas Eve. It seems unlikely that flying saucers could be seen so often, and never leave unambiguous evidence behind.

A common response from UFOlogists is that the evidence exists, but is being covered up and hidden by government. It seems that about 15 per cent of Americans are deeply suspicious of government. A recent example is the belief that the Apollo moon landings did not take place but were faked in a movie studio. This belief was spread largely through the Internet, and through a lurid "documentary" broadcast on TV. The website http://www.badastronomy.com has been very helpful in providing antidotes for such claims. See Philip Plait's book *Bad Astronomy*.

The so-called evidence for these claims includes images of US flags that are apparently waving, rather than drooping, even though there is no wind on the moon. It's instructive to show students the actual pictures of this and other purported evidence, and to explain how NASA propped up the lunar flag on purpose, so they can see for themselves.

Another related pseudoscience is the claim that a specific feature on the surface of Mars, first imaged by the first Viking orbiter, is actually the image of a face, created by a long-lost Martian civilization. One of us (JRP), who happens to be a fan of *The Muppet Show* (a popular show from the 1980s for children, young and old), counters this claim by showing students an image, from Venus, of a volcanic feature, called Aine, which looks remarkably like one of the stars of *The Muppet Show* – Miss Piggy. This is appropriate, since all of the features on Venus, with one exception, are named after famous women. Further, Malin Space Science Systems, which built the Mars spacecraft that has recently taken high-resolution observations from orbit, has shown on a webpage (http://www.msss.com/education/facepage/face.html) how detailed views of the "face on Mars" feature show how ordinary it is.

References

Carlson, S. 1985, *Nature*, **318**, 419.
Comins, N. 2001, 2003, *Heavenly Errors: An Explanation of Misconceptions About the Real Nature of the Universe*, Columbia University Press, hardback 2001, paperback 2003.
De Robertis, M. M. and Delaney, P. A. 1993, *Journal of the Royal Astronomical Society of Canada*, **87**, 34.
De Robertis, M. M. and Delaney, P. A. 2000, *Journal of the Royal Astronomical Society of Canada*, **94**, 112.

Pasachoff, J. M., Cohen, Richard J., and Pasachoff, Nancy W. 1970, "Belief in the supernatural among Harvard and West African students," *Nature*, **227**, 971–2.

Plait, Philip, 2002, *Bad Astronomy: Misconceptions and Misuses Revealed, from Astrology to the Moon Landing "Hoax,"* New York: John Wiley.

Richardson, W. Mark and Wildman, Wesley, eds., 1996, *Religion and Science: History, Method, Dialogue*, New York: Routledge.

Shermer, Michael, 2002, "The gradual illumination of the Mind," *Scientific American*, **286**(2), February, 35.

Wynn-Williams, G. 1998, "A very quick statistical test of newspaper astrology predictions," in A. Fraknoi, ed., *Proceedings of the Symposium on Teaching Astronomy for Non-science Majors*, San Francisco: Astronomical Society of the Pacific.

Editors' Note: Pasachoff's course *Science and Pseudoscience*, taught at Williams College in 2003 and repeated to a much larger attendance in 2005, deals with a wide variety of pseudosciences, misuse of statistics, and misleading beliefs. The syllabus and bibliography are online at http://www.williams.edu/Astronomy/courses.

Part VI

Astronomy and culture

Introduction

Astronomy is deeply rooted in almost every culture, as a result of its practical applications and philosophical implications. Nowadays, we determine time, date, and direction from clocks, calendars, compasses, and global positioning system (GPS) signals from the sky. In earlier times, these were determined by direct observation of the sky. Even today, the Islamic month and new year are based on direct observation, not on a calendar prepared in advance. One of us (JRP) lives and teaches in the most multicultural city in the world – Toronto. There, a large fraction of students come from Asian cultures that set their calendar by the sun and/or moon. The other (JMP) comes from another multicultural city, New York, where the lunar Jewish calendar still resonates in a large fraction of the population. The Christian calendar also approved by and named after Pope Gregory XIII in 1582, now often called 1582 CE (Common Era) but formerly AD (Anno Domini) 1582, also has many astronomical connections, and students may be interested to learn about these.

The connection between astronomy and religion, of course, can be problematic. In Part V, we included creationism under the heading of pseudoscience. But astronomy and religion need not conflict. Our astronomical colleagues in the Vatican Observatory are both astronomers and Jesuits. Over the years, they have developed a deep understanding of the possible relationships between science and religion, while maintaining a first-class astronomy research program (and a major collection of meteorites). Nevertheless, many humans are deeply influenced by their belief and faith in their religions, as well as by their culture.

Many issues can be introduced through the history of astronomy. This study provides an excellent opportunity to teach about the nature of science and of scientific discovery. It demonstrates that science is a human endeavor. And it adds human interest! After all, Copernicus, Tycho, Kepler, Galileo, and Newton were interesting characters, as were Einstein and Hubble in more recent times. Today, students may be fascinated by characters like Steven Hawking and the late Carl Sagan.

The names that we have mentioned are obviously European, but other cultures have contributed substantially as well. Islamic astronomers nurtured and developed astronomy during the Dark Ages. For centuries, East Asian astronomers kept systematic records of comets and novae. Pacific Islanders including the Maori used astronomical observations for practical purposes. Planetariums have regularly been used to show how these people navigated by the stars. In Chapter 14, Julieta Fierro eloquently describes ancient structures in Mexico that had astronomical significance.

Astronomy is all around us. For lucky students in cities such as Rome, astronomy is embedded in ancient buildings, as Leonarda Fucili described in Chapter 6. For students in "younger" countries such as the USA and Canada, students must be content with the astronomy contained in popular culture – advertising images, rock music, and commercial product names such as Mercury, Saturn, and Subaru automobiles, Pulsar watches, and Quasar TVs.

14

Teaching astronomy in other cultures: archeoastronomy

Julieta Fierro

Mexico's National University, Apdo. Postal 70–264, 04510, Mexico DF

Abstract: In this paper I shall discuss the way in which an archeology lecture can be used to explain some astronomical topics so that Latin American students will relate better to them. More than 2,000 years ago great constructions were used for naked-eye observations and timekeeping in Central America. The zenith pass was fundamental for the calendar; teachers can use this event to explain the measurement of the Earth's circumference. All great cultures have had astronomers, and many countries can use their past to make modern science more interesting and meaningful for their population.

14.1   Introduction

Mesoamerica is a large zone that includes most of Mexico and Central America. In that region one of the earliest written languages was invented. A large civilization flourished thanks to the discovery of agriculture and the use of a mixture of corn, beans, squash and peppers that conveys a high nutritional content. In this part of the world there are mainly two seasons, the dry and rainy periods. The slave-driven civilization that developed in this part of the world consequently divided the civil year in two, mainly dedicated to agriculture and construction. A good calendar was needed for central planning and commerce.

14.2   The sun

One of the main deities of ancient Mexico was the sun. It had several representations, including an eagle that drifts through the sky, and the "Ollin," which means movement. The eagle is depicted in Mexico's flag, and the "Ollin" can clearly be seen in the center of the Aztec Calendar and the ten peso coin (see Fig. 14.2). A jaguar was the symbol for night, with the spots representing stars. Warriors dressed like eagles and jaguars.

In ancient Mexico, the sun's trajectory was well studied. The Mesoamerican people carefully registered the places where the sun rises and sets during the year. At mountain-surrounded sites they used the silhouette, or carved vertical slots to mark the exact location of the rising sun (Fig. 14.4). At flat locations, they built special constructions to register such spots. This way the calendar based on the rising and setting sun was established throughout the region. Mesoamericans also had a 260-day ritual calendar, which ran in parallel to the astronomical one.

Some of the markers for special events such as the equinoxes are the ballgame courts as seen in Fig. 14.5. This game was played with a large rubber ball, using the hip to make it go through a stone loop. On the days of the equinoxes, the sun sets precisely in the direction of the ball court, and can be seen through the stone loop.

181

Fig. 14.1. Ancient Mexico.

Fig. 14.2. Center of the Aztec Calendar reproduced on a coin.

There were other constructions built to mark this day, such as the "Castle" at Chichén-Itzá, where the sun forms shadows showing the descent of a snake along the pyramid's stairs (Fig. 14.6). The teacher can bring a souvenir pyramid to the classroom and light it with a lamp in order to simulate the sun's movement and show the way the Maya people could build scale models of their constructions and orient them in a way that they could be useful for astronomical purposes.

Another marker for the equinox in the Maya region is the "Caracol," often said to be an ancient astronomical observatory. The name Caracol means snail. This rounded construction

Fig. 14.3. Ek Balam, where a warrior dressed like an eagle is represented in full scale.

has a spiral shaped hall with several windows that are aligned with particular celestial events such as the rising of Venus on determined dates of calendric importance.

14.3   The moon

The Aztecs believed the dark areas of the moon represented a rabbit; they depicted the moon as a hare in a pot. In fact the moon's diameter is the same length as Mexico, as seen in Fig. 14.7. In Náhuatl, the language of the Aztecs that is still spoken in Mexico, Mexico means "the moon's navel." Teachers may use the hare image to talk about lunar cratering and volcanism.

Once the moon has been addressed, the teacher can explain eclipses. He or she can begin by talking about traditions that are still in order since Aztec times. For instance, during total eclipses pregnant women tie red ribbons around plants and scissors around their necks "in order to avoid having a deformed child" (in pre-Hispanic times an obsidian arrow head was used instead of scissors). Teachers may explain that the creation of present Mexico City was at the time of a total solar eclipse. Tenochtitlán was founded on a lake where an eagle (the sun) was resting on a cactus plant (this image is part of the nation's flag).

The way in which pre-Hispanic people predicted eclipses by carefully observing the sun's and moon's trajectories can be explained to students by telling them that, since the paths of the sun and moon form a $5°$ angle, and since their apparent motion is different, with the moon moving slower, one can infer when the trajectories will cross (Fig. 14.8). It is important to point out that the moon's rising is more erratic than the sun's and that this jumping of the moon's motion was another reason for attributing the hare as its deity. It is also an invitation for students to observe its motion.

Part of the rituals included fights among eagle days and jaguar nights, which symbolized the sun and the night.

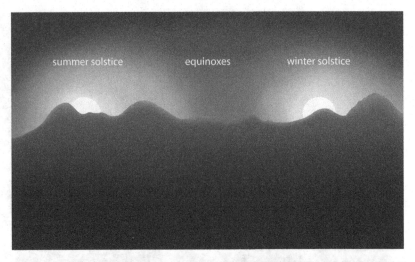

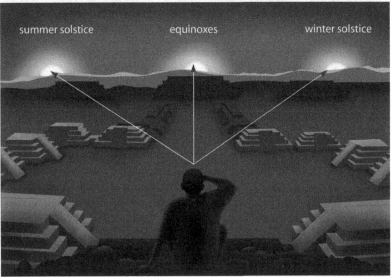

Fig. 14.4. Rising and setting of the sun using mountains and pyramids to measure the location.

14.4   Measuring the Earth's circumference during a zenith pass

One of the most prevailing characteristics in Mesoamerican constructions is a perforation drilled in order to observe the zenith pass of the sun. The ancient calendar began in May with the first rainfall. The zenith pass occurs during that same period. I shall give a few examples. At Xochicalco, there is a cavern built inside one of the pyramids on the main square, where the sun illuminates it during the zenith pass (Fig. 14.9). Incidentally, an astronomical convention was held there during the fifteenth century. People from all over Mesoamerica came to decide on calendar matters, which were important for commerce.

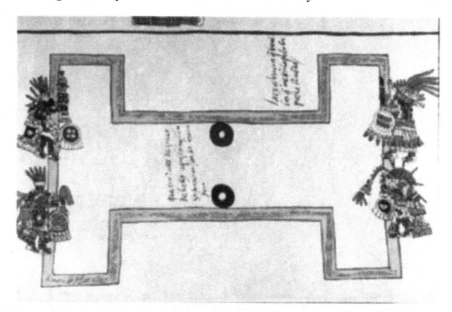

Fig. 14.5. Ball court from a drawing in the Codex Borbonicus.

Fig. 14.6. (Left) Scale model that simulates a serpent's shadow on the staircase of the Castle Pyramid when illuminated from the correct angle; and (Right) the actual view of the Castle Pyramid at Chichén-Itzá.

At Malinalco, a monolithic pyramid is sculptured from a rock mountain, and the sun illuminates a sculpture of an eagle, the sun representation, during the zenith pass (Fig. 14.10). At Monte Albán there is a small construction with a different orientation than the rest of the site, whose staircase points to the pyramid dedicated to the zenith pass. At the same site (and others) there are several gnomons.

A teacher from this area can use this historical fact to show how these sites that are at different latitudes could be used to measure the circumference of the Earth "à la Eratosthenes." Using a flat strip of foam rubber, the teacher can simulate a flat Earth with obelisks at the same longitude but different latitudes. He or she can use the same strip to simulate perforations, that is to say, zenith passes. Eratosthenes used an obelisk and a well; the Mesoamericans could have used perforations. A very good assignment is to use the Internet to have students

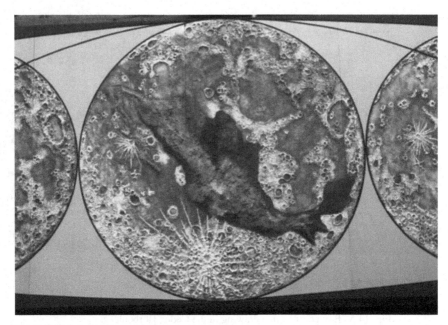

Fig. 14.7. Relative sizes of Mexico and the moon.

Fig. 14.8. (Left) Simulation of the sun and moon trajectories to explain that they cross in the sky. (Right) The Dresden Codex, where eclipse predictions are pointed out.

at different locations measure the circumference of the Earth by observing on which day the zenith pass occurs at their location.

14.5　Angular measurements

The ancient people from Mesoamerica were stargazers. Stars were symbolized as eyes. They invented a series of constellations. For instance Orion was represented as a grinding stone, and the Big Dipper (an asterism, not a constellation, to North Americans) by a kite. They used two long rods to measure the altitude of stars above the horizon. The angular distance was

Fig. 14.9. Xochicalco and the effect of the sun's rays penetrating into a cave.

Fig. 14.10. At Malinalco (right) the sun illuminates an eagle (left) during the zenith pass.

determined by pointing one of the rods towards a marker placed on a ceremonial construction and the other at the celestial object.

This fact gives the teacher a chance to address angular measurements on the celestial sphere and to describe the way an astrolabe can be used to determine the altitude of an object above the horizon. A very simple astrolabe can be built using a compass and a weight.

It is often difficult for students to use maps to locate constellations, mainly because of the small scale, because the aspect of the sky varies during the course of a night and also

Fig. 14.11. The teacher can use a rubber strip with rods to explain the way shadows at different latitudes are an indication of the Earth's circumference.

Fig. 14.12. Pre-Hispanic representation of an astronomer, and the way two rods were used to measure angular distances. Notice the stars represented by eyes around the construction.

throughout the year, and because all the stars shown on the map are not necessarily visible in the sky due to light pollution, etc.

In order to avoid these problems the teacher can draw a couple of constellations on rigid plastic sheets with fluorescent paint or with fingernail polish that glows in the dark. The correct scale is about 30-cm for Orion. Local as well as the conventional constellations can

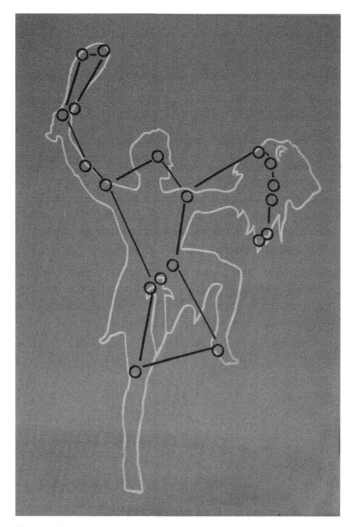

Fig. 14.13. A rigid, transparent plastic sheet, with a drawing of stars and the outline of a constellation, that can be used to find that constellation.

be used. The teacher should only draw as many stars as are visible in his or her particular location. When the arm that is holding the sheet is outstretched, the size of the constellation in the sky should be the same as the one on the drawing.

14.6   Conclusion

Astronomy is a fascinating topic. Its multidisciplinary nature makes it teachable in several ways, including those that make more sense to a particular group of students. In countries that have great historical tradition, it can be employed to help pupils relate to science. In places such as Mexico, teachers can take pupils to one of the sites, and explain naked-eye observations, and from there build on to modern astrophysics. This experience will enrich the learning process.

Comments

Harry Shipman: Is there any evidence that Mesoamericans used the stars for navigational purposes?

Julieta Fierro: I'm not sure – I'm trained only concerning archeoastronomy and the calendars.

Nick Lomb: Do you take students to the archeological sites that you mentioned with astronomical/calendrical connections?

Julieta Fierro: Yes – the students have great pride in their past culture.

Rosa M. Ros: Congratulations on your presentation. It is an excellent idea to introduce activities related to the cultural background of your country. It is possible to introduce some examples in order to connect astronomy with buildings that exist in a lot of cities, buildings that were oriented according to astronomical knowledge. It is important to introduce these examples in the classroom because the students know these examples very well. This is an excellent opportunity to connect astronomy with the real world.

Editors' Note: Bradley Schaefer (2004), schaefer@lsu.edu, has evaluated several reported archaeoastronomical sites, and finds it improbable that the Caracol at Chichén-Itzá was used for astronomy. His comments are to appear in the Oxford VI Conference (International Conference on Archaeoastronomy and Astronomy in Culture).

Poster highlights

The sole poster in Part VI is entitled **Introducing astronomy through solar and lunar calenders** and comes from **Moedji Raharto**.

In Indonesia, the lack of competence of many teachers in basic science, astronomy, and space science implies that knowledge of astronomy and space science will be transmitted to the young generation improperly.

Priority in a curriculum of basic science includes only a small amount of general astronomy. The public perception is that astronomy is less important than basic science. Both of these points create a disadvantage for the developing astronomical community in Indonesia, a country with more than 230 million people.

The Muslim community in Indonesia has a tradition of using the lunar calendar to determine the first day of the important months Ramadhan, Syawal and Dzulhijjah. Recent disputes over determining the first day of these months is partly due to a lack of understanding of how the exact time of the first visibility of the lunar crescent is calculated by astronomers.

The challenges of introducing astronomy to a wider community with little background concerning astronomical education was discussed in this paper.

Part VII

Astronomy in developing countries

Introduction

One of the International Astronomical Union's highest priorities is the worldwide development of astronomy. A substantial fraction of its resources is used for that purpose; the resources are administered through the IAU's Commission on Education and Development. At the 2000 General Assembly of the IAU in Manchester UK there was a three-day conference on "Astronomy for Developing Countries." The proceedings were edited by Alan H. Batten, and published by the IAU in 2001. This book is the definitive guide to the topic. The IAU works closely with other organizations, such as the United Nations Office for Outer Space Affairs. This office has organized a series of workshops on basic space science, in various parts of the world, and astronomy is one of the topics covered. The government of Japan has generously provided small telescopes and planetariums to several developing countries. The Vatican Observatory organizes summer schools for graduate students and young astronomers from astronomically developing countries. The IAU International Schools for Young Astronomers (ISYA) take place every year or so somewhere in the world, and also involve graduate students and young astronomers. See www.astronomyeducation.org.

According to the *StarGuides* database maintained at the Strasbourg Observatory, there are approximately 100 countries in which there is some astronomical activity – either research or teaching, professional or amateur. In about 50 of these countries, astronomical activity is not sufficiently developed for the country to adhere to the International Astronomical Union (IAU), or there may be other factors that make IAU membership difficult. Their level of astronomy activity ranges from minimal to moderate. There may be amateur astronomy activity and/or teaching of astronomy in the schools, but the membership and activity of the IAU is primarily restricted to post-secondary and professional astronomy. Incidentally, there is no international organization of amateur astronomers, or of astronomy educators. Of the approximately 50 countries that adhere to the IAU, about 25 have significant difficulty supporting astronomy because of economic reasons; these include countries that have only recently emerged from the "third world," and countries of Eastern Europe, the Baltic, and the former Soviet Union, whose economies are still in a precarious condition. The situation with regard to astronomy research and education in these countries is highly variable; some may discontinue astronomical activity and IAU membership (though we hope not); others may soon recover or progress to a higher level. The remaining 25 countries are well developed, and have economies that are strong to reasonably strong (though, in some, scientific research and education are facing serious restructuring and cutbacks, with serious repercussions for astronomy).

The term "astronomically developing country," of course, is relative. Even countries such as the USA have regions that could be described as astronomically undeveloped!

In Part VII, we present two interesting case studies. In the first, Case Rijsdijk describes recent efforts to improve the teaching of astronomy in South Africa. In that country, there is an active professional astronomy community. With the dismantling of apartheid, that community has worked actively to spread astronomical teaching and research to the populations that were not previously served. Then Jay White describes the IAU's projects in Vietnam, a country which, after three decades of war, is working to restore its astronomical activity. A key goal of the IAU's involvement is to train a new generation of astronomy teachers for secondary schools and universities; a unique textbook in the Vietnamese language is a cornerstone of that project.

See also Chapter 21 entitled "A short overview of astronomical education carried out by the IAU," by Syuzo Isobe, concerning the activities of the IAU. His paper includes a list of the Commission on Education and Development's International Schools for Young Astronomers (ISYAs).

Throughout this book, we have also included summaries of contributed papers from our colleagues in the astronomically developing countries. In many cases, they were not able to attend the conference for economic reasons. Their accomplishments are remarkable, especially considering the challenges they face.

15

Astronomy curriculum for developing countries

Case Rijsdijk

"Masintinge," 600 Friedheim, Highstead Road, Rondebosch 7700, South Africa

Abstract: The educational needs of countries differ: this is as true in developing countries as it is in developed countries and applies to all subjects taught. Despite these different needs, there are fundamental ideas, concepts, topics and threads of commonality within the curricula of all subjects, including specialized subjects such as astronomy, which can be exploited in the creation of new curricula for developing countries. The recent experience in South Africa of developing a new Outcomes-Based Natural Sciences Curriculum, which includes some astronomy, could well prove to be a useful framework, or starting point, for the creation of an astronomy curriculum for other developing countries, and possibly even developed countries. Since the subject "astronomy" is often seen as elitist, or an unnecessary luxury, several ways of integrating or "hiding" astronomy within existing curricula will also be considered.

15.1   Introduction

South Africa's recent history in education has been fairly turbulent as a result of previous government policies, especially in how it affected the education of black people. After the democratization of South African society, it became clear that there was a need to develop a new curriculum, and it was an ideal time to start with a "clean slate." South Africa had the unique opportunity to make a major paradigm shift in education and make a total break with past structures. One of the aims of this new curriculum was to transform education. This transformation would be not only educational but also political: there was a need to empower previously disadvantaged (black) teachers. This process started in the early 1990s, and the implementation of a new curriculum began in 1998. The new curriculum had a distinct South African flavor, since it had to be relevant in a local context, though a similar approach could be used by other countries. Since astronomers (and other scientists) have little, if any, educational experience at school level, the South African efforts at transformation also give an indication of how astronomers, and other scientists, can best intervene to maximize their impact and make a real contribution to the development of a new curriculum by using their expertise and existing resources, in addition to those given by the Internet. In developing countries, the Internet is not always accessible, so new and alternative resources need to be developed, which could well be adapted by developed nations.

15.2   Background

Prior to the new curriculum, South Africa had a schools curriculum for each population group in each province – there was no national curriculum, although in certain cases there was a lot of similarity among various curricula. In all there were 19 different departments looking

197

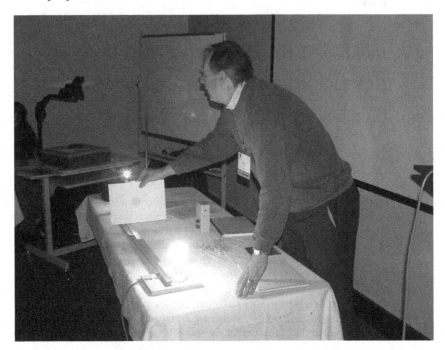

Fig. 15.1. Case Rijsdijk (South Africa) demonstrating the properties of light at a teachers' workshop in Sydney, July 26, 2003. Photo by Rob Hollow.

after South African education. In addition each population group[1] had its own schools with widely differing facilities, resources, and teaching capacity. White students had access to the best facilities and teachers, while their black counterparts, under the Department of Education and Training (DET), had access to the poorest facilities and under- or unqualified teachers.

This was particularly true in the sciences: often the only exposure black students had to science was biology. What physics and chemistry they learned is what they picked up in the course of learning biology and other life sciences. Astronomy formed a part of geography and, since very few geography teachers had any knowledge of science or astronomy, the subject was sometimes "done" in one classroom session, but usually omitted completely. It was also perceived to be "elitist" and difficult and only of use to "gifted" children. In developing a new curriculum, the South African Department of Education consulted curriculum developers from New Zealand, Australia, Scotland, the USA, Canada, and the Netherlands. The resulting curriculum was an outcomes-based education structure, similar to that used in Australia and New Zealand, and was known as "Curriculum 2005," C2005 (DoE 1997). It represented a revolutionary change in our educational system. C2005 was implemented in 1998, and shortly afterwards it became clear that there were implementation problems. It was a complex document using a structure and language that was unfamiliar to most educators who had not had the time to be properly prepared for such a major change. A review panel was set up and

---

[1] South Africa in the past (apartheid government) recognized the following as distinct population groups: whites (those of European descent), colored (those of mixed parentage), Indian (descendants from the Indian subcontinent), and black (the indigenous peoples of South Africa).

headed by Professor Linda Chisholm. Her report was submitted to the government in 2000 (Chisholm 2000) and resulted in the new Revised National Curriculum Statement, RNCS (DoE 2002). This curriculum is currently being implemented and will be constantly reviewed and updated. Its structure has been greatly simplified with fewer, more specific outcomes, and the language has also been made more "user-friendly" and not quite as academic as it was previously.

15.3   Curriculum 2005

Curriculum 2005 is an Outcomes-Based Education (OBE) curriculum that has two bands: the General Education and Training, GET, band and the Further Education and Training, FET, band. The GET band covers grades R (preschool)–9 and the FET band covers grades 10–12. The FET band is in the process of being finalized and will be implemented in 2006. These higher grades will continue with their "traditional syllabus" until then, and so the focus in this paper will be on the GET band.

Here the traditional "subjects" have been replaced by eight "Learning Areas" (LAs):

- Language and Literacy;
- Mathematics and Mathematical Literacy;
- Arts and Culture;
- Natural Sciences;
- Human and Social Sciences;
- Technology;
- Life Orientation;
- Economic and Management Sciences.

The GET band is divided into three phases: Foundation (Grades R–3), Intermediate (Grades 4–6), and Senior (Grades 7–9). For the Foundation phase, the LAs are clustered into three: Literacy, Numeracy, and Life Skills. The eight Learning Areas are divided among these three, and the Natural Sciences Learning Area is subdivided into four "Themes:"

- Matter and Materials (ex. Chemistry);
- Life and Living (ex. Biology);
- Change and Energy (ex. Physics);
- Earth and Beyond (ex. Geography).

These themes replace the traditional subjects. The "Earth and Beyond" was originally further divided into:

- Beneath the Earth;
- On the Earth;
- Above the Earth;
- Beyond the Earth.

Here the "Beyond" component was, of course, "astronomy," which covers the traditional introductory topics such as lunar phases, seasons, time and time zones, eclipses, the solar system, stars, galaxies, and the evolution of the universe. There are many positives to this new curriculum, such as the fact that it is learner-oriented[2] in that the learner is expected to

---

[2] The word learner is used in South Africa because people of greatly differing ages are studying the same materials. The word "pupil" had colonial connotations, and in South Africa "students" are those at tertiary institutions.

do the work: the teacher is seen as the "guide on the side" rather than the "sage on stage." The emphasis is on the learner "doing things" rather than "rote learning" and "chalk and talk," as was the case in the past. Assessment is more varied and multifaceted: learners are expected to "show" understanding through projects and other work, which is continuously assessed. Should a learner not be able to "show" satisfactory understanding, the learner is given additional work and further opportunities to "show" understanding.

Grade 9 is also seen as an exit point from "formal" education: it is anticipated that many learners will not continue onto the Further Education and Training (FET) band for a variety of reasons, including a lack of funds and the need to go out to work to support the family unit.[3] To accommodate these people, the new curriculum encourages a philosophy of "life-long learning." The idea here is that when people enter the workplace they are encouraged to continue learning and obtain more skills and further qualifications. Another important consideration about the RNCS is that it is a National Curriculum. This feature allows learners to freely move around the country without being disadvantaged as a result of differing curricula. In addition it means that tertiary institutions do not have the problem of trying to evaluate competencies submitted on behalf of learners who come from different backgrounds or provinces.

15.4   The new curriculum and some South African solutions

When such a large paradigm shift takes place, much is expected from teachers, and there are bound to be implementation problems. These problems are mentioned in some detail, as it seems likely that the South African experience may well be repeated in other developing countries embarking on a new curriculum. Many of the problems encountered are exacerbated by the fact that the bulk of our teachers, who are poorly trained and underqualified (Rijsdijk, 2001), are now expected to deal with totally different lesson and assessment strategies, in addition to using new textbooks and other unfamiliar learning materials. To accommodate this lack of expertise in the RNCS, many teachers are expected to attend after-hours workshops on how to:

- implement continuous assessment in the classroom;
- use new resources;
- develop new learning programs.

An additional problem was that geography teachers were now expected to teach some science as well: the subject geography straddles two Learning Areas, Natural Science and Human and Social Science. Many teachers were reluctant, as these teachers often took geography at their respective training college only as a second subject. As a result, in many schools the "Beyond" part of "Earth and Beyond" was expected to be taught by geography teachers who had found a home in the Learning Area "Human and Social Sciences." This meant that astronomy became a part of the social sciences and was often taught by teachers with little or no science knowledge or training. As was mentioned earlier, astronomy was perceived to be difficult and elitist anyway, so it was just ignored.

---

[3] The situation is exacerbated by the AIDS pandemic: many children are now orphans, and the eldest child needs to support his/her siblings. Another often overlooked fact is that many teachers are HIV+, and there are already signs that this fact is affecting the teaching body in South Africa.

There are specific problems when a totally new subject/topic is added to the curriculum, particularly when the subject, such as astronomy, is perceived by many educators as unnecessary and of no practical value – rather like Latin! So having successfully negotiated the inclusion of astronomy, it now became imperative to develop new resources that were relevant and showed teachers that astronomy was useful and did have a place in the classroom. There was some resistance: astronomy was difficult, and nearly all teachers thought it had to be done at night. It was a "specialist" subject suitable only for really clever people or experts. So for teachers who were already under strain attending workshops in learning how to cope with the Revised New Standard Curriculum, the addition of extra workshops to learn how to use these newly developed and supplied resources proved challenging!

Another interesting problem was the division of grades across different levels of school. In South Africa, primary (or junior) school goes from preschool to grade 7, and high school, or senior school, from grade 8 to grade 12. Thus the senior phase of the GET band (grade 7 to grade 9) falls partly in primary school and partly in high school. This division is serious, as children go from being at the top of the school to freshers at the bottom of the high school during a crucial phase of their learning process. They are also exposed to very different teaching philosophies and strategies in the high school, often using very different books, with the result that material may either be repeated or fall between the "cracks."

15.5   Alternative strategies

In 1994 the author did a survey of educational structures and performance in Zimbabwe (Rijsdijk, 1994). Both Zimbabwe and South Africa had similar educational structures before Zimbabwean independence in 1980, and it was felt that a comparison of the South African situation in 1994 with that of Zimbabwe after 14 years of independence might give some indication of a "best practice" approach. After independence, Zimbabwe adopted an "evolutionary change" approach: they left what was usable in place and changed only what needed changing. Further changes took place in parallel with teacher training, alleviating the implementation problems that South Africa is at present experiencing. The survey showed that evolutionary change is better than revolutionary change, since it was found that the Zimbabwean education system was better, in all respects, than the South African one, and consistently produced better results at all levels.[4]

The strategy should be to keep what does not need changing, especially in mathematics and science, but to change the curricula of subjects such as language, history and the social aspects of geography. Where change is needed in other subjects, it must be done in an incremental way. Teachers should be consulted in the process. They will be empowered by involvement: make them part of the process, make it their curriculum. It would probably be a good idea to start by looking at the assessment process as we have found that this impacts critically on the development of the curriculum.

15.6   Intervention by scientists

Few scientists are also educators, particularly at the school level, although astronomers are probably in a slightly better position than other scientists in this regard, because of the nature

---

[4] Political instability in Zimbabwe in recent years has impacted severely on the educational system, which, like many other aspects of life there, is experiencing problems, so the above comparison is probably no longer valid. Additional factors include the AIDS pandemic (see previous footnote).

Fig. 15.2. Case Rijsdijk with the winners of the prize for "making the best model of an astronomical event" at the annual Science Festival, SASOL SciFest.

of the subject. To tap into this vast pool of expertise, at the school level, and to maximize its impact, the South African experience has shown that the best option is for scientists to:

- assist curriculum developers with content and ideas;
- collaborate with textbook authors;
- work with teachers in developing resources;

- work with teachers at training institutions;
- liaise with local education departments and officials.

Other interventions are possible even where no curriculum change is taking place. In any existing chemistry, physics, or mathematics curriculum there is ample material that can be used, either as illustrative examples or as "add-ons" in topics such as:

- the inverse square law;
- spectroscopy;
- optics;
- gravitation;
- nuclear physics;
- magnitudes;
- graphs, etc.

Additional interventions are possible by supplying data and ideas to repeat simple experiments and to give support material for project work.

15.7   Resources

The Internet is a wonderful resource, but it is not one that is readily available to many people in developing countries. It is either unavailable or, where it is available, it is expensive or available only in a very limited way: one computer per school served with a 54-kb modem and telephone line.

In addition, much material is written from a "northern perspective" and needs to be adapted; much is also overly complex in language and content. Often sites are located that contain poor or bad material, and local teachers have insufficient knowledge to identify these sites. Another interesting problem is that learners would rather believe NASA than their teachers! There have been cases where learners used NASA material that was intended for the northern hemisphere. When the teacher said this was wrong, the learners simply didn't believe her! After all, NASA can't be wrong! There was a positive outcome in that, when it was explained to them why it was wrong, the learners accepted the explanation (eventually), and their respect for the teacher soared. The issue dealt with lunar phases and it might be an idea for these sites to add in something to explain that there is this north/south difference.

There is a "digital divide," and it is widening. Developing countries experience problems with:

- bandwidth;
- hardware;
- software;
- support and maintenance;
- servers.

The Internet does have its place, but it is not the magic solution: if you want to show a child a box, then give that child a box, and the child will immediately *know* what a box is. No amount of work on a computer screen will give that child comparable knowledge of that box. The author is collaborating with others in using the Internet to develop these tactile resources, *and* other resources for use by teachers and learners (Rijsdijk, 2000; 2001; 2003). What is important in a developing country is that the resources use readily available materials, and are cheap and easily reproducible.

15.8   Conclusion

In creating a new school curriculum in a developing country, it is worthwhile to consider the following:

- see what other developing countries have done: what works in the USA or Australia might not work in a developing country;
- consider the context in which the new curriculum is being developed; this is especially important when inviting outside advisors;
- use local, appropriate, and relevant material: take cognizance of Indigenous Knowledge Systems, IKS;
- create or write new material, including textbooks, although this is expensive and time-consuming. Often it is the publishers that drive the curriculum, since they are the ones producing the textbooks. This process can also lead to a skewed implementation process (Viall, 2003);
- look at assessment strategies;
- work closely with teachers and local education authorities;
- start the process on *completion* of the new curriculum!

The school curriculum is important, and scientists, especially astronomers, should be contributing to that. But it must be remembered that practicing teachers have often settled into "comfort zones" and are reluctant to change. In-service training is necessary and good, and many NGEO's and National Facilities have active Education and Public Outreach (EPO) programmes that assist with these. But we should be looking at the curriculum in the teacher training colleges. If we provide these students good exposure to and experience in using astronomy as a teaching aid, then, irrespective of the curriculum, the teachers know that astronomy can be used as a "vehicle for science education" (Rijsdijk, 2000). Then they will know when, and how, they can use astronomy in their teaching of science and mathematics: astronomy becomes the comfort zone. While this may not be classed as "astronomy education," it does lay a sound groundwork for future education. In South Africa's case, there is more astronomy in the FET band, and if the ground work has been done in the GET band, even if under a different guise, it will lead to a better knowledge of astronomy.

Education is about human interaction. In South Africa we have the anti-Cartesian view, which we call "Ubuntu," or "Motho ke Motho ka Batho:" I am because we are, and because we are (therefore) I am. It is teachers who interact with the class; computers, textbooks, etc. are tools used by people: by themselves they have only limited value. Astronomers need to support and interact with teachers, especially during their initial training period.

15.9   Acknowledgments

I would like to express my thanks to the IAU and the National Research Foundation, NRF, for making my visit to the 25th General Assembly possible. In addition I would like to thank the staff at the South African Astronomy Observatory (SAAO) for continued support and to all the teachers who spent time in helping to develop and test resources.

References

Chisholm, L. 2000, "Report of the Review Committee on Curriculum 2005."
Department of Education (DoE) 1997, "National Curriculum Statement, Curriculum 2005."
Department of Education (DoE) 2002, "Revised National Curriculum Statement, Grades R–9."
Rijsdijk, C. 1994, "Report commissioned by the Diocesan College," Bishops, Cape Town: Unpublished.

Rijsdijk, C. 2000, "Using Astronomy as a Vehicle for Science Education," *Proceedings of the Combined Conference of the ASA and RASNZ*, Sydney 2000: Proceedings of the Astronomical Society of Australia, **17**, 156.
www.atnf.csiro.au/pasa.

Rijsdijk, C. 2001, Initiatives in astronomy education in South Africa. In A. Batten, ed., *Astronomy for Developing Countries, IAU Special Session at the 24th General Assembly*, San Francisco: Astronomical Society of the Pacific, p. 117.

Rijsdijk, C. 2003, "Doing it without electrons," in A. Heck and C. Madsen, ed., *Astronomy Communication, Astrophysics and Space Science Library*, **290**, Berlin: Kluwer Academic.

Viall, Jeanne 2003, "Education gets wrong outcome," *Cape Argus* (newspaper), November 19, 2003.

Comments

Julieta Fierro: How do you deal with different South African languages for astronomy education?

Case Rijsdijk: South Africa has 11 different languages, so English is the language that is officially used by the state, and in education. Zulu is the most widely spoken indigenous language, followed closely by Xhosa. There is at present a vigorous debate about what languages the textbooks should be written in. There are other problems, such as the level of vocabulary. An example: the word "amandla" means force, power, energy, and work! Using different languages is also expensive, and eventually they'll use English. A possible solution would be to use English textbooks, but to use other languages in discussions and explanations.

16

Science education resources for the developing countries

James C. White II

Rhodes College, Department of Physics, Memphis, Tennessee 38112 USA

Abstract: Considerable attention is paid to how we in the so-called developed countries teach science and to how our students learn it. Even though similar questions are asked by scientists and science teachers in all countries, one finds that resources available for science education in developing countries are often scarce but not unobtainable.

16.1   Introduction

Resources for science and science education are often quite limited in countries less wealthy than those of, say, North America or the European Union. Leaving aside for the moment those resources for scientific research – equipment for experimentation, computer hardware and software for theoretical and/or data analyses, communication infrastructure (journals, Internet access, etc.), and opportunity for collaboration with scientists outside the country – one confronts the needs of a country for the development and advancement of its science education system: equipment for experiments and classroom demonstrations, computer hardware and software for simulations and/or data analysis, communication infrastructure (textbooks, Internet access, etc.), and collaboration with science teachers inside and outside the country. That the needs of the scientific enterprise and those of the education enterprise are so similar is not surprising, given that good science can enrich teaching, and engaging teaching can compel one, student or teacher alike, to ask new questions and, hence, to conduct new, enriched science.

16.2   Education needs

To many policy makers in wealthy and comparatively poor countries alike, enriched science education necessarily means increased funding, which, in the zero-sum game of politics, means less money for other national needs. Further, crafters of policy quite often have difficulty dissociating the needs for science, as in *research*, from those of education.

Indeed, a painful refrain still heard occasionally is that developing countries have no need for basic science and, hence, emphasis on science and science education is diluted or completely washed away. In response to a suggestion concerning the viability of theoretical physics as an area of research in developing countries, a delegate to a forum of the International Atomic Energy Agency remarked: "Theoretical physics is the Rolls Royce of sciences – what the developing countries want is nothing more than bullock carts" (Salam, 1987). This delegate, an economist, considered his field the primary concern of governments in developing countries, with science merely one of the bells and whistles of the development process. What he failed to realize, however, is the impact that even simple programs of science can have to

206

invigorate the development process and, most importantly, intellectually marshal a country's students, the next generation of its leaders.

It is incorrect to assume that expensive science, as in *research* demanding great national expenditures, is necessary to improve the quality of science education. The best science experiences, it turns out, often emerge from the simplest and most inexpensive activities.

16.2.1 Equipment

If one walks into a physical- or natural-science classroom in the United States, one will discover lab benches scattered with scientific instruments (air tracks, force probes, oscilloscopes, etc.) and shelves filled with scientific apparatus. The same is not typically true for a science classroom or lab space in a developing country.

Despite the seemingly endemic electronic devices that whirl and buzz and spit out data to an associated computer, teachers in relatively wealthy countries around the world are now realizing that some of the most effective demonstrations and laboratory experiences for our students come from simple, inexpensive sources. As examples, one can

- use a burning teabag to demonstrate convection;
- have students investigate air resistance with falling feathers, strips of paper, pebbles, and a wristwatch;
- make analogy between thumping a melon for ripeness and divining Earth's interior with the use of seismic knocks; or even
- use a rotating desk chair to discuss conservation of angular momentum.

Such simple activities with basic, available materials place student and teacher alike fully in the domain of hands-on learning. Equally important is that, once the student leaves the classroom, she can locate the same materials to reproduce the experiment or activity inexpensively for herself, again and again.

Rather than considering how best for us to fund equipment purchases for science classrooms in developing countries, we should use "equipment" available locally in the country – teabags, bits of twine, melons, pebbles, and so forth. All we need do is guide teachers there in its use, making sure that use of food or other specific items does not offend local sensibilities or customs, and in how interpretation of results can enrich the classroom experience for their students.

16.2.2 Communication

Educational research and common sense suggest that effective communication is a critical element in student learning – communication between the teacher and student, among students, and among teachers as they discuss what works and does not work in their classrooms. The Internet is clearly an effective means of bringing teachers and pupils together in non-traditional means – for example, having electronic office hours after the school is closed for the day – but one can not assume that the relatively easy Internet access one has in, for example, the United States, is available in developing countries. Moreover, should a science educator in a developing country have Internet access, the cost for that access can be prohibitive.

For this reason, standard means of communication, verbal and written, must still be relied upon for the majority of learning in the developing world. We astronomy educators in wealthier nations can turn to print publications like *The Physics Teacher* or the Astronomical Society of the Pacific's *Universe in the Classroom*, publications that often have electronic, web-based versions, or to electronic-only "publications" like the new *Astronomy Education Review* (http://aer.noao.edu). Yet in developing countries textbooks are often the main source of information.

When a science teacher selects a textbook, he or she must consider factors such as price to the students or the school system, whether the book will be available in time for the class or in the numbers needed, the language used (e.g., is a specific book written in English available also in Spanish?), the topics covered, and the like. For a teacher in a developing country, each factor is more pronounced: the cost of a single copy of the ideal book is often greater than the teacher's monthly salary, exclusive of shipping; translations into the vernacular are not always available; and should a teacher find a donor of suitable books, the "freshness dating" on those books is long expired. Science is quite dynamic and quickly evolving, and although fundamentals in a given field may be more static, every teacher wishes to provide her or his students with current information.

Recognizing the importance of textbooks, particularly in developing countries, a segment of the IAU's "Teaching Astronomy for Development" Program Group (part of Commission 46) worked in 1999–2001 with Vietnamese astronomers and educators to write and then publish a new astrophysics textbook. The undertaking is unique to this point in time, and the textbook created just happens to incorporate recommendations made at IAU Colloquium 105, on *The Teaching of Astronomy*, in 1988, concerning textbooks for teachers in developing countries: "[a] jointly written text [is] more desirable, as this will allow the local author to inject the local bias of the book" (Othman, 1990).

The first Vietnamese textbook to contain color illustrations, *Astrophysics*, was written by four Vietnamese astronomers and one American astronomer. Both Vietnamese and English are used in the text, with the Vietnamese language on left pages facing English translations on the right (see pages 210, 211). Such a design presents students with the science and also provides them with opportunity to refine their English language skills and improve their English science vocabulary.

16.2.3   Collaboration

Discovering what does and does not work in someone else's classroom can save one time and effort. Yet fostering communication among science educators in developing countries and with their colleagues in wealthier countries is difficult: technological, cultural, and language barriers stand ready to make such dialogue difficult, if not to prevent it entirely.

The IAU's Commission 46 facilitates better communication through various activities, and the "Teaching Astronomy for Development" Program Group (TAD) is charged explicitly with offering assistance, advice, and guidance to science educators in those developing countries with little astronomy that wish to enhance their astronomy education (and science education, in general) significantly. From visits to the countries by foreign scientists to assisting financially select students from developing countries with travel and associated expenses for their graduate educations abroad, TAD seeks to minimize the loneliness a scientist or science educator feels in a country with few resources for education and science.

By offering advice or services that the host country believes it needs, TAD, and the other, allied Program Groups in Commission 46, attempt to connect each scientist and science educator with the rest of the world – bringing the individual, and his or her students, into the global scientific endeavor.

References

Othman, M. 1990, "Textbooks: a panel discussion," in J. R. Percy and J. M. Pasachoff, eds., *The Teaching of Astronomy*, IAU Colloquium 105, Cambridge: Cambridge University Press, 1990.
Salam, A. 1987, In C. H. Lai, ed., *Ideals and Realities*, Singapore: World Scientific, 12.

Comments

Bill MacIntyre: How much cross-cultural communication with the developing country did you receive and consider when developing the textbooks/resources? I am asking this because there are cultures where using food for anything other than eating is taboo.

James White: We don't go into another culture without communicating with the host country, and they will be able to inform us about what is appropriate with regards to textbooks and resource development.

Moedyi Raharto: I am interested in the program to give support to write astronomical textbooks in local languages (Indonesian) because the government gives priority to support basic sciences (excluding astronomy), and we need such books for students' general reference. How many programs of writing astronomical textbooks in local languages do you have?

John Percy: The IAU can support only a handful of countries through its Teaching for Astronomical Development (TAD) program. Vietnam was one of those countries.

Bruce Partridge: What role can scientists from a developing country, but now working abroad, play? Certainly in the case of Vietnam, there are many distinguished scientists working in France and elsewhere.

Michèle Gerbaldi: Vietnamese people who are living in France have kept strong ties with Vietnam, and amongst them there are several French professional astronomers. For example, one of them, Dr. Nguyen Quang Rieu, collaborated with Dr. Donat Wentzel to produce the first modern astrophysics manual in Vietnam, a bilingual edition in Vietnamese and English. Also, Vietnam students did (and are doing) Ph.D. theses in astronomy and astrophysics in France using grants. It is important that cooperation exists with various countries and international organizations such as the IAU.

C. R. Chambliss: What was the approximate cost of the Vietnamese astrophysics book that you mentioned? Is it affordable in Vietnam? I believe that Indonesia has published vernacular technical literature (in Bahasa Indonesia) more than any other Asian nation.

Jay Pasachoff: The IAU/TAD spent about $3500 on the Vietnamese textbook.

Chương VII.B

BÊN TRONG MẶT TRỜI

Làm thế nào để ước tính độ sâu bên trong Mặt Trời?

Chúng ta không thể quan sát Mặt Trời bằng ánh sáng khả kiến. Nhưng lí thuyết cho chúng ta rất nhiều thông tin. Chúng ta bắt đầu với sự thật rằng phần bên trong của Mặt Trời ổn định trong hàng triệu năm. Mặt Trời phải ở trạng thái cân bằng thủy tĩnh. Hãy tưởng tượng Mặt Trời được chia thành các lớp cầu. Khí trong mỗi lớp chịu tác động của lực hấp dẫn kéo nó xuống. Để giữ lớp khí ở độ cao không đổi, áp suất khí ở phía dưới lớp phải cao hơn so với ở phía trên lớp. Độ chênh lệch cần thiết về áp suất p(r) ngang theo một lớp có độ dày dr, ở cách tâm một khoảng cách r là:

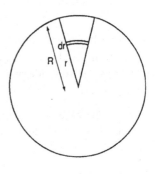

Hình VII.17

$$dp = -g(r)\rho(r)dr \qquad (7.1)$$

Trong đó g(r) là gia tốc hấp dẫn và $\rho(r)$ là mật độ khí. Vế phải là lực hấp dẫn tác động lên chất khí trong một xi lanh nhỏ có tiết diện ngang 1 m^2 và độ dài dr. Vế trái là sự chênh lệch về áp suất (lực trên một đơn vị diện tích) tại đỉnh và tại đáy của xi-lanh.

Gia tốc hấp dẫn g(r) có giá trị như khi tất cả khối lượng M(r) bên trong quả cầu bán kính r được đặt tại tâm quả cầu, $g(r) = -GM(r)/r^2$. (Chúng ta có thể chứng minh được điều này bằng một phép tích phân đơn giản để tính lực hút hấp dẫn giữa một mẫu khí và tất cả các mẫu còn lại trong ngôi sao, hoặc chúng ta cũng có thể chứng minh điều này một cách khá đẹp đẽ hơn bằng cách sử dụng hệ phương trình vi phân). Khối lượng bên trong một lớp khí hình cầu có độ dày dr là dM(r) = $4\pi r^2\rho(r)dr$.

Chúng ta có thể biểu diễn dp từ phương trình VII.1 dưới dạng:

$$dp = \frac{GM(r)}{4\pi r^2}dM(r) \qquad (7.2)$$

Phương trình này tiện ích hơn phương trình 7.1 vì nó được diễn tả chỉ theo một hàm số chưa biết M(r), và chúng ta biết giá trị của nó tại bề mặt, M(r=R) = M, khối lượng đã biết của Mặt Trời. Để tính tích phân một cách chính xác phương trình 7.2, chúng ta cần sử dụng thêm nhiều kiến thức vật lí. Nhưng chúng ta cũng đã có thể biết được một số thông tin quan trọng về bên trong Mặt Trời.

Chúng ta có thể chứng tỏ một cách khá chặt chẽ rằng áp suất tại tâm của một ngôi sao như Mặt Trời phải rất lớn. Chúng ta biết rằng dp/dr < 0 ở khắp nơi để duy trì trạng thái cân bằng thủy tĩnh. Bởi vậy, áp suất tại tâm p(tại tâm) phải là áp suất lớn nhất trong ngôi sao, và nó phải lớn hơn bất cứ áp suất nào được tính trung bình cho toàn Mặt Trời. Ở đây chúng ta chọn một giá trị trung bình của p(r) theo toàn bộ khối lượng Mặt Trời

$$\langle p \rangle = \int p(r)\frac{dM(r)}{M} = -\int M(r)\frac{dp(r)}{M} \qquad (7.3)$$

Fig. 16.1. A page in Vietnamese from the Vietnamese textbook, *Astrophysics*. When looking at a spread of facing pages in *Astrophysics*, students see the text in Vietnamese on the left-hand side, and on the right-hand side they see the English version of the text (Fig. 16.2). The mathematics is preserved in the translation.

Chapter VIIB

THE SOLAR INTERIOR

How to estimate the deep interior of the Sun.

Figure VII.17

We cannot look into the Sun using visible light. But theory gives a lot of information. We start with the fact that the interior has been steady for millions of years. The Sun must be in hydrostatic equilibrium. Imagine the Sun divided into spherical layers. The gas in each layer experiences gravity pulling it down. To keep the layer at a constant height, the gas pressure must be higher below the layer than above it. The needed difference in pressure, p(r), across a layer of thickness dr at distance r from the center is

$$dp = -g(r)\rho(r)\,dr\,. \tag{7.1}$$

Here g(r) is the gravitational acceleration and $\rho(r)$ is the density of the gas. The right side is the gravitational force acting on the gas in a small cylinder with $1\,m^2$ cross section and length dr. The left side is the difference in pressure (force per unit area) at the top and bottom of the cylinder.

The gravity g(r) is the same as if all the mass inside the sphere of radius r , M(r), is placed at the center of the sphere, $g(r) = -GM(r)/r^2$. (One can show this either by a rather tedious integration of the gravitational attraction between one piece of gas and all the other pieces of gas in the star, or one can show it rather elegantly using partial differential equations). The mass inside a spherical layer of thickness dr is $dM(r) = 4\pi r^2\,\rho(r)\,dr$.

We now can express dp from equation (7.1) as :

$$dp = \frac{GM(r)}{4\pi r^2}dM(r) \tag{7.2}$$

This form is more useful than equation (7.1) because it is expressed in terms of just one unknown function, M(r), and we know its value at the surface, M(r=R) = M, the known mass of the Sun. For a precise integration of equation (7.2), more physics is needed. But already one can learn some significant information about the solar interior.

We can show quite rigorously that the pressure at the center of a star like the Sun must be very high. We know that dp/dr < 0 everywhere in order to maintain hydrostatic equilibrium. Therefore, the central pressure, p(center) must be the largest pressure in the star, and it must be greater than any pressure averaged over the Sun. We choose here an average of p(r) over the mass in the Sun,

$$\langle p \rangle = \int p(r)\frac{dM(r)}{M} = -\int M(r)\frac{dp(r)}{M} \tag{7.3}$$

Fig. 16.2. A page in English corresponding to that in Vietnamese in Fig. 16.1 from the Vietnamese textbook, *Astrophysics*. On the right-hand side of a spread of facing pages in *Astrophysics*, students have available to them the English translation of the Vietnamese text. Having two languages in front of them simultaneously permits students to exercise not only their scientific skills, but also their language skills.

Part VIII

Public outreach in astronomy

Introduction

The conference on which this book is based focused on astronomy education in the schools, where our society passes on knowledge and understanding to the younger generation in a systematic and formal way. In the proceedings of an astronomy education conference held in Maryland in 1996, however, Andrew Fraknoi wrote as follows:

Let me begin by posing the following question: where does astronomy education take place in the United States? Those readers who teach will probably say that it takes place in classrooms like theirs, anywhere from first grade through university. But I want to argue that astronomy education happens in many other places besides the formal classroom. It happens in hundreds of planetariums and museums around the country; it happens at meetings of amateur astronomy groups; it happens when someone reads a newspaper or in front of television and radio sets; it happens while someone is engrossed in a popular book on astronomy, or leafs through a magazine such as *Sky and Telescope*; it happens in youth groups taking an overnight hike and learning about the stars; and it happens when someone surfs the astronomy resources on the Internet. When we consider astronomy education, its triumphs and tribulations, we must be sure that we don't focus too narrowly on academia and omit the many places that it can and does happen outside the classroom.

Astronomers, their institutions, and their organizations actively promote understanding of astronomy through "outreach" in all of these ways. In the USA, such outreach has recently been formalized by NASA's requirement that major projects have Education and Public Outreach (E/PO) activities.

Outreach is a rather nebulous term that means different things to different people. It may refer to "informal education" of students, outside the classroom. It may refer to "science promotion" through "public relations" and glossy brochures. It may refer to public awareness, understanding, and appreciation of science – three very different issues. Science museums and planetariums play a major role in such outreach. It may refer to the development of general science literacy in society. Science literacy is promoted by governments on the grounds that it is necessary for a healthy economy. Outreach can take many forms: school visits by scientists; professional development, activities, and resources for teachers; websites for teachers and students; activities at science centers; science clubs, in and out of school; science camps, and other on-campus programs at colleges and universities; programs targeted at under-served groups such as women and Aboriginal people; science fairs and other contests; research projects for students (and teachers); field trips; job shadowing and other career information; public lectures and displays (and in the case of astronomy, star parties); books, magazines, TV programs, and other audiovisual materials; and awards for outstanding science activities.

Outreach is sometimes equated with public education only, but outreach to teachers is especially important, given their usual general lack of knowledge about astronomy and astronomy teaching. The Astronomical Society of the Pacific's *Project ASTRO* is exemplary in this regard. It partners professional and amateur astronomers with teachers, not just for a one-off lecture, but for an ongoing relationship. The *Project ASTRO* manual emphasizes that the astronomer and the teacher bring complementary skills to the relationship: one is an astronomy expert, and the other is a teaching expert.

In the USA, agencies such as the National Science Foundation (NSF), and the National Aeronautics and Space Administration (NASA), promote science education and outreach, and support it through grants. There are "umbrella organizations" such as the American Association for the Advancement of Science that also speak for science, and science education in general, for instance, through sessions at their annual conference each February; see also www.aaas.org.

In astronomy, planetariums, science centers, and public observatories play a special role. These are highlighted in Chapter 18 by Nick Lomb and in a poster paper by Tony Fairall. They also raise the important question of how the planetarium community can best be linked to astronomical organizations such as the IAU. Astronomy journalists and writers were not explicitly represented at this conference. Sometimes a professional astronomer, in this conference, Mexico's Julieta Fierro, takes on the additional role of science journalist and writer; she has won many international awards for this work. The important contributions of amateur astronomers to astronomy education were also not considered explicitly at this conference. Those contributions are remarkable in their quality, quantity, and variety – and are completely voluntary. In Canada in 2003, the (mostly amateur) Royal Astronomical Society of Canada won the Michael Smith Award, the top national award for science outreach.

Astronomers, concerned about the need to communicate more effectively with the public, recently drafted the "Washington Charter" – an appeal for and guide to effective communication and outreach. It is reproduced and discussed later in this book. The Charter was a culmination of a pair of meetings on "Communicating Astronomy" held in Tenerife and in Washington, DC, respectively. The first was run in 2002 by Terry Mahoney of the Astrophysics Institute of the Canary Islands, the second was run in 2003 by Charles Blue of the US National Radio Astronomy Observatory, and the third was run by Lars Christensen at the European Southern Observatory near Munich: http://www.communicatingastronomy.org.

There are a number of significant issues in astronomical outreach, and some of the programs, mentioned above, specifically address them. Outreach programs should reach the under-served, not just the elite. They should reach families and communities; there is a limit to what can be accomplished in schools. Outreach should be more than public relations; it should inform and educate ("steak, not just sizzle," as the saying goes). At the same time, it should convey excitement and enthusiasm. It should, if possible, attract students to science careers, broadly defined. It should make science part of our culture.

There are effective way to achieve this goal. One is to concentrate on teachers or, better still, teacher educators ("train the trainers") or, even better still, those who produce the curriculum, the textbooks, and other high-leverage material. No one individual or organization can do this; partnerships and networks are essential, both at the national and the local level. And the effectiveness of programs should constantly be assessed. Are they meeting their goals? Are they serving the users?

17

What makes informal education programs successful?

Nahide Craig and Isabel Hawkins

*University of California, Berkeley Center for Science Education,
Space Sciences Laboratory Berkeley, Ca 94720–7450, USA*

"Total Solar Eclipse 2001" – live from Africa

Abstract: Evaluation and assessments of informal education programs have been challenging because of the diverse nature of objectives, setups, and expected outcomes of these programs. Almost all institutions that develop and present such programs include evaluation specialists in their staff. However, for very large public outreach efforts, large evaluation groups/institutions can contribute more objective and extensive evaluation and assessment instruments that will help to identify whether the program was successful and if the learning objectives were achieved. Approximately 42,000 people participated in "Total Solar Eclipse 2001" at 164 public venues, including 21 museums internationally, science centers, and planetariums. We will expand further on the properties of this program that help to determine whether it was successful or not. Success can be affected by such issues as personal interest in the content, publicity, connectivity to a group, educational as well as entertainment value, and the challenges of using high technology.

17.1 Introduction

Informal education can be defined as the overlap between formal education (i.e., K–14 curriculum development, educator workshops, and links to systemic reforms) and public outreach (i.e., Internet, popular science articles, educational TV, radio programs). Informal education combines educational substance with public outreach, but without the pressure of examinations and assessment. More explicitly, it includes museum exhibits, science center programs, and planetarium shows. It can include educational activities carried out by community organizations such as scouts, girls and boys clubs, 4H, and other youth groups. It engages students, educators, and the general public in settings away from the classroom; provides learning opportunities, and motivates further learning and lifelong interest (Morrow, 2000).

Science centers are considered to be the best venues to bring informal education to the public. They inspire interest in science and scientific exploration, often with hands-on or activity-based science programs. These programs increase students' creativity, positive attitudes toward science, perception, logic development, communication skills, and reading readiness. These programs also encourage interest in careers in science, engineering, technology, and influence students' career path. Science centers also have close ties to the media and are often called upon to explain recent news, providing wide dissemination of their programs and high visibility. In addition, the attendance statistics at science centers, obtained by the Association of Science and Technology Centers (ASTC), indicate that more than 60 per cent of the adult public in the United States go to science centers and museums at least once a

217

year, and that 22 million people visit about 1,400 planetariums within the USA. Also, the fact that science centers work directly with students through school outreach programs and field trips, reaching an estimated 39 million school children every year, makes the science centers the best locations for informal education, and thus the best potential partners for informal education programmers.

Evaluation and assessments of informal education programs, small or large, such as science museum traveling exhibits, interpretive kiosks, hands-on activities and very large public programs, have been challenging because of the diverse nature of objectives, setups, and expected outcomes of these programs. Almost all institutions that develop and present informal education programs include evaluation specialists in their staff. However, for very large public outreach efforts, which include participation of many institutions located across the country, larger evaluation groups/institutions can contribute more objective and extensive evaluation and assessment instruments. Such instruments will help to identify whether the program was successful and if the learning objectives were achieved. They can also lead to "lessons learned" for future events and serve as possible model evaluation instruments for informal education institutions such as science museums and science centers where the budgets do not allow for contracting independent reviewers.

17.2   "Eclipse 2001" and lessons learned

The "Total Solar Eclipse 2001" event was developed and executed with the partnership of NASA's Sun–Earth Connection Education Forum (SECEF), The Exploratorium (the Museum of Science, Art and Human Perception, in San Francisco), and NASA's STEREO Mission (a future solar space mission). American Institutes for Research (AIR), an independent evaluation company from Boston, was contracted to develop and implement the evaluation. The following informal educational goals were used to guide the event:

- uses a "hook" to highlight science and engage the public;
- creates an experience where visitors learn and retain scientific knowledge;
- inspires interest in science and scientific exploration;
- provides wide dissemination and high visibility of national scope.

Approximately 42,000 people participated in the "Eclipse 2001" event at museums, science centers, and planetariums; a total of 164 institutions worldwide, including 80 scientists, 71 Girl Scout troops, 15 universities and schools, and 2 mass-media outlets (CNN and NASA TV). AIR developed evaluation forms, which were sent to all the participating organizations; 972 evaluations from public participants were received and analyzed. In addition, 37 organizations and 32 scientists sent in their evaluations. The participants reported learning more about concepts such as sunspots, the solar corona, the diamond ring effect, and general information about the sun. Several institutions mentioned frustration with technical difficulties but, overall, respondents in all categories reported that they enjoyed (a) the broadcast of the eclipse; (b) the hands-on activities; and (c) interaction and online chats with the scientists. The majority of children and adults indicated that they would participate in a similar event in the future, and evaluations demonstrated a statistically significant increase in their knowledge about the sun after the eclipse.

The participating institutions wanted their sites to host the next total eclipse event of 2006. (*Editors' Note*: See also the IAU Commission's website at www.eclipses.info). The scientists reported that they were able to discuss their research with participants, that hosting museums and the Sun–Earth Connection Forum had prepared them well before the event, and that they will participate again in a future event. Scientists also mentioned that they were motivated to participate because they could turn kids on to science, talk to the public, have the opportunity to discuss their research, and also enjoyed watching the "Eclipse 2001" web-cast.

From the "lessons learned" aspect we can suggest the following guidelines for a successful informal education event:

- scout out candidate science centers;
- conduct front-end studies that focus on visitors' interests; take this prior knowledge into account when designing the program/exhibit;
- plan for formative and summative evaluations;
- plan workshops for the docents, volunteers, and educators;
- schedule public presentations for participating scientists;
- get your website ready for background information;
- prepare printed interpretive materials;
- consider technical difficulties for the new technologies.

Looking back at our goals, we feel confident that the "Eclipse 2001" event, where we could not see the event from our locations but could "observe" it remotely, was the "hook" to highlight science and engage the public for this event. We demonstrated that we created an experience where visitors indicated they learned and that this event inspired interest in science and scientific exploration. Through the science museum connections, we provided wide dissemination and high visibility of national scope. The personal interest of the scientists in the content, publicity, connectivity to a group, educational value, and the capabilities of using high technology defined the success.

17.3 Future events

Encouraged by this event we plan other informal education programs, involving many science centers and planetariums across the country and the globe. The first such program was an interactive video conference focusing on our new knowledge about how our universe began. This event was developed by the Adler Planetarium and is entitled "Journey to the Beginning of Time." It took place on October 15, 2003. On June 8, 2004, a celestial event of historical scientific importance occurred, when the silhouette of the planet Venus crossed in front of the sun as seen from the Earth. The last transit of Venus occurred in 1882, so no one alive today had ever witnessed the transit of Venus. An exciting live web-cast presentation of this event, "Venus Transit," took place on June 8, 2004, from Spain. An "Ancient Observatories" live web-cast took place on March 20, 2005, from Chichén Itzá, Mexico, with the Exploratorium broadcasting it live. Their 2005 program also included Chaco Canyon, New Mexico, and Hovenweep, Utah. The next "Total Solar Eclipse" live web-cast is also planned on March 29, 2006, at a location on the path of totality yet to be determined. We hope you will participate in some or all of these programs. For information please visit our website: http://solarevents.org.

References and further reading

Morrow, C. A. 2000, "A Framework for Developing Education and Public Outreach Programs Associated with Scientific research Programs," unpublished.
Association of Science and Technology Centers (ASTC), http://www.astc.org/resource/case/index.htm
National Aeronautics and Space Administration (NASA), 2004, http://sunearth.gsfc.nasa.gov/sunearthday/2004/index_vthome.htm

Comments

Julieta Fierro: Linking science centers by the Internet during special events such as eclipses or openings is a wonderful experience and worth the effort. The next world conference on informal education will be held in Brazil, and is sponsored by the Network for the Popularization of Science and Technology in Latin America and the Caribbean Association for Science Outreach (Red-POP; Red de Popularización de la Ciencia y la Tecnología para América Latina y el Caribe).

John Percy: Check out www.redpop.org. The conference was in April 2005 in Rio de Janeiro, Brazil. Previous to the Science Centre World Congress in Rio there is a meeting of the Network for the Popularization of Science & Technology in Latin America and the Caribbean.

18

The role of science centers and planetariums

Nick Lomb

Sydney Observatory/Powerhouse Museum, PO Box K346, Haymarket, NSW 1238, Australia

Abstract: The school curriculum in many countries includes astronomical topics such as the seasons, phases of the moon, planets, and stars. Yet teachers at all school levels generally do not know much astronomy and have difficulty teaching that part of the curriculum. Even if they have some knowledge of the subject, they may not have the resources to illustrate it and to create enthusiasm in their students. One solution is to take them to a place specializing in astronomy education – a suitable science center or museum or planetarium or public observatory.

18.1   What are science centers and planetariums?

Science centers and planetariums are places that are dedicated to illustrating and explaining astronomical concepts. There are different types of institutions, though some have elements of more than one:

- Science centers have interactive or "hands-on" exhibits. They cover a variety of scientific subjects that in some cases include astronomy.
- Planetariums project star fields and astronomical images on a curved dome above an audience.
- Museums have objects and displays. Like science centers, they cover a variety of subjects; in some cases they include astronomy.
- Public observatories have telescopes that are available to the public.

18.2   Why take students to a science center or planetarium?

Teachers take their students to science centers, planetariums, or similar places for a variety of reasons:

- **Instruction**: students can be instructed by someone knowledgeable about astronomy.
- **Stimulation**: students will be stimulated by the exhibits, the show and the ambience.
- **Enjoyment**: students will enjoy the experience.

A comment from a teacher illustrates the incentive for going on excursions: "The kids get really excited and so do I, when it is somewhere that will be special and show us things that we can't see or do anywhere else" (Quadrant Research Services, 1998). However, teachers may face a number of negative considerations that hinder them from going to these places:

- **Cost**: it could be expensive to reach the science center. For example, it may be necessary to hire a bus.
- **Missed classes**: teachers of other subjects may not want the students to miss their classes and so may oppose long excursions.

- **Fear of working outside normal hours**: an evening visit to a public observatory is out of normal working hours for teachers.
- **Difficult questions**: students may ask questions after a visit that the teacher may not be able to answer. This is a situation a teacher may find embarrassing and so may want to avoid.

The Powerhouse Museum has found (Whitty, 1999) that links to school curriculum are not always reason for an excursion, but that they are important for the teacher to justify it. Teachers consider that a visit must meet not only academic goals but also social, cultural, and vocational ones. They value the opportunity of experiential learning for their students in contrast to the normal verbal experience of the classroom, yet at the same time they want them "to see something real."

18.3   What do students want?
The Powerhouse Museum recently completed a "front-end evaluation" for a new Space Exhibition (Stollznow Research, 2003). The evaluation took the form of a series of representative members of the public being invited to discuss their attitudes to the subject in small groups. One group was made up of 14- to 15-year-old lower-secondary students. Some results from that group were as follows:

- **Interests**: the students were interested only in what directly involved them. For example, they were interested in what it is like living in space, but only because they thought that they may have the opportunity to go there in the future.
- **Interaction**: they want exhibits with interaction and still more interaction.
- **Experiences**: they would like exhibits that allow them to experience weightlessness (rather difficult to provide!), go in a spacecraft, etc.
- **Best exhibit**: the best exhibit that the students experienced was where they could ride a bike and see how fast they had to ride to make something such as a hair dryer work.

18.4   What happens during a visit?
We take as examples visits to Sydney Observatory, which is a combination of a public observatory and museum, and two planetariums, one a medium-sized one in a large city, the Melbourne Planetarium, and the other a relatively small one in a small city, the Launceston Planetarium.

Sydney Observatory
The Sydney Observatory is the oldest existing observatory in Australia and is now, as part of the Powerhouse Museum, a museum of astronomy and a public observatory. It has over 80,000 visitors a year, of whom around 20,000 are school students. It has displays of historic astronomical instruments, modern interactive exhibits, a small planetarium and a 3D theater. School visits are highly structured with the students divided into groups of 15 or so, each accompanied by a guide-lecturer. During the $1\frac{1}{2}$ hour visit the guide-lecturer concentrates on the topic, such as the solar system or the nature of stars, requested by the teacher in advance.

During the visit a variety of very different experiences is provided to the students. They can look through the two large telescopes in the building at the sun (with appropriate filters), or at a star if it is clear, or at something terrestrial near the horizon if it is cloudy. They visit the exhibition, concentrating on areas dealing with the solar system if they are primary students, as that is generally requested by their teachers. They also visit the small planetarium and the 3D theater, where they wear polarized glasses to go on a trip to Mars or on a journey around the solar system. Most important of all, during their visit to Sydney Observatory they have the opportunity to interact with a guide-lecturer and to ask him or her lots of questions about the planets and the stars. Teachers often indicate their appreciation of the facilities provided to students. For many groups the time spent in the new 3D theater is especially exciting. Follow-up materials for the teachers usually include a copy of the annual *Sydney Observatory Sky Guide*, notes on the solar system, and the Observatory's astronomy time-line poster.

A new kit called *Sydney Observatory on the Move* has been developed for country schools that are too geographically distant from Sydney to visit the Observatory. This kit is to be sent to a regional area, where schools will be able to share it among themselves. It contains a range of astronomy materials that are unavailable in most schools, especially in regional areas. The materials include a small telescope, a sun, Earth and moon model, spectroscopes, posters, videos and books. There is also a detailed teacher's guide in the kit with examples, curriculum links, and information.

Melbourne Planetarium

The Melbourne Planetarium was established in its present form in 1999. It is located within the Scienceworks Museum, which is a campus of the Museum of Victoria. It has a 16 m dome with 150 unidirectional seats surrounding a Digistar II digital planetarium projector. Of the planetarium's 100,000 annual visitors, about 40,000 are school students. A variety of planetarium shows is offered for students of different ages, including:

- **Tycho Stars Again**: Discovering the wonder of the stars and telling stories about the constellations. For ages 5+ (Preschool–Year 3).
- **Cosmic Couriers**: Discovering just how big the universe really is. For ages 7+ (Years 3–6).
- **Spinning Out**: What makes the sun rise? Why do we have seasons? For ages 7+ (Years 3–4, 7–8).
- **Launch Pad**: Exposing all the weird but true facts about the solar system. For ages 8+ (Years 4–8).
- **Guiding Lights – Navigating by the Stars**: How explorers, from the English captain James Cook to the Apollo astronauts, have relied on the stars to find their way. For ages 10+ (Years 5–12).
- **Escape from Andraxus**: Uncovering the amazing diversity of stars within the universe. For ages 12+ (Years 7–10).

The shows are pre-recorded, but they are introduced by a planetarium staff member who also presents a live 15-minute show on what is in the sky. Teachers receive a comprehensive education kit related to the show that includes curriculum links, activities, resources, and a show summary.

A recent survey (Scienceworks, 2002) indicates that the planetarium is viewed positively by schools. In July 2002, the Scienceworks Museum carried out a survey of schools to gauge their satisfaction with visiting the museum and the planetarium. There were 253 returns of questionnaires from schools, representing a 10 per cent response. With regard to the

Table 18.1. *The cognitive domain*

Category	Description	Sample learning objectives
Remembering	Recall facts	List the names of the nine planets; distinguish daytime and night time
Comprehension	Simple understanding	Recognize the difference between a star and a planet; identify the Earth's spin as the cause of day and night
Application	Use a concept in new situation	Classify the sun as a planet or a star; explain why stars appear to move around the celestial poles
Analysis	Pull apart information into its component parts	Compare the conditions on the surfaces of the inner planets; derive the direction of the Earth's spin from the rising and setting of the sun and the stars
Synthesis	Put facts together to form new meaning	Find the position of a star at a given date and time with a planisphere; deduce why stars rise earlier each day
Evaluation	Make judgments about concepts	Link the spin direction of the planets with the formation of the solar system; evaluate the calendar with its 12 months and seven-day week, and propose a new one

planetarium, 97 per cent of responders indicated that the shows held student interest, 93 per cent that the shows had entertainment value, and 95 per cent that they had educational value.

Launceston Planetarium

Established in 1967, the Launceston Planetarium is part of the Queen Victoria Museum in Tasmania. It has an 8-meter dome with 45 concentric seats. There are about 8,000 visitors per year, including 4,000 school children. Though these numbers may seem small compared with a planetarium in a major city like Melbourne, they represent a significant percentage of the local population of 70,000.

For school students, the planetarium shows are aligned to the school curriculum. There are separate shows for lower primary and upper primary, while high-school students see the public shows. Each show consists of three separate segments: a pre-recorded segment of 20 to 25 minutes, a live commentary on the night sky, and question time.

Teachers receive handouts relating to their visit, including a sheet of most asked questions and answers. Informal feedback on the planetarium visits is positive as illustrated by the same teachers bringing back their new crops of students each year.

18.5   An educational framework

Visits to museums and planetariums should take into account how students learn. According to the standard model of learning developed by Benjamin Bloom (1956), there are three types or domains of learning. These are *cognitive*, which relates to knowledge and mental skills, *affective*, which relates to feelings and attitudes, and *psychomotor*, which deals with manual and physical skills. Learning objectives for programs, whether exhibition visits or visits to planetarium shows, will necessarily favor those from the cognitive domain, but should also try to include ones from the other two as well.

The cognitive domain can be subdivided into categories. These are listed in Table 18.1 together with some sample learning objectives. Many museum visits or planetarium shows refer only to the lowest category of remembering, but for a meaningful experience the visit should incorporate a range of cognitive-learning categories.

18.6   Does learning take place?

Museums and planetariums can provide an informal learning environment to students that is very different to the formal one in a school. In a museum setting, students come across ideas in random sequence, and they can make their own choices of what they look at in depth. They learn in an unstructured fashion by experimenting, by observing, and by talking, discussing, and collaborating with their peers.

The museum environment is deliberately designed to allow informal learning where the visitor is an active participant in the process. Museums follow a constructivist learning approach according to which "learning is a development process involving the accommodation of new experiences with prior understandings and attitudes" (Griffin, 1999a).

Unfortunately, teachers are often afraid to allow their students to use this informal environment fully. Griffin (1999b) found that many teachers impose a structure on the students' experiences. They control which exhibits they are allowed to see, they control what video-based exhibits they can watch, and they insist on the filling in of worksheets. These worksheets impose an order into what the students look at and prevent full exploration of hands-on and other displays.

According to Weber (2002), "The emphasis on education has shifted from abstract to concrete experiences, while the process of knowing has become more important than the accumulation of knowledge." Museums not only have an informal environment, but also provide concrete experiences. Seeing a planetarium show is a concrete experience that students cannot have elsewhere. Seeing "real objects" such as a historic telescope is a concrete experience, as is looking through a telescope or using a well-designed hands-on exhibit.

Children are likely to remember these experiences better than the facts that they are illustrating. Assessing what learning takes place under these informal conditions is difficult and may not be productive. The very personal nature of learning in a museum, the short time students are involved in these distinct experiences, and the broader, but individual contexts that occur make it meaningless to attempt to measure museum-based learning with the same degree of reliability as classroom learning (Griffin, 1999a).

Instead of testing cognitive learning from a museum visit, Griffin suggests looking at behaviors indicative of learning such as "making links and transferring ideas and skills" and "sharing learning with peers and experts." In a video-recorded study of a school group visiting the Australian Museum, she found that these indicators provide a useful method of determining if the conditions for learning are present or not.

18.7   Discussion

Science centers, planetariums, museums, and public observatories provide the interaction and variety of experiences that both teachers and students value on school excursions. They assist teachers who may not be fully knowledgeable about astronomy to communicate the subject to their students. Museums (science centers, planetariums), however, are not efficient places for traditional "school-type" education, that is, for learning specific facts and concepts (Whitty, 1999). The students do not spend enough time on an excursion for that purpose and are usually too excited to be in a receptive state of mind. Instead, museums and planetariums are ideal places for providing wonder, for the opportunity of exploring a variety of concepts and for expanding young minds.

18.8   Acknowledgments

I would like to acknowledge valuable discussions and assistance from Helen Whitty of the Powerhouse Museum's Education and Visitor Services, and from Carol Scott of the Museum's Evaluation and Audience Research Unit. Martin George from Launceston Planetarium, and Tanya Hill from Melbourne Planetarium, were also most helpful. My colleagues from Sydney Observatory, Toner Stevenson and Jeanie Kitchener, provided useful information and assistance.

References

Bloom, Benjamin, ed., 1956, "Taxonomy of educational objectives, the classification of educational goals," in *Handbook 1: Cognitive Domain*, New York: D. McKay, 62–197.

Griffin, Janette 1999a, "Finding evidence of learning in museum settings," in E. Scanlon, E. Whitelegg, and S. Yates, eds., *Communicating Science: Contexts and Channels*, London: The Open University with Routledge, 110–19.

Griffin, Janette 1999b, "Formal education groups in informal settings: helping teachers to find an effective balance," presented at "Challenge of Change in Education" symposium, University of Technology, Sydney, Kuring-gai Campus.

Quadrant Research Services 1998, "Evaluation of Educational Programs and Services," prepared for the Powerhouse Museum, unpublished, 53.

Scienceworks 2002, Schools Booking Database Survey, Report No. 263, unpublished.

Stollznow Research 2003, "Market Research Report on Permanent Exhibition on Space: Front-end Evaluation," unpublished.

Weber, Traudel 2002, "Museums and schools: a review of the relationship," in M. Xanthoudaki, ed., *A Place to Discover: Teaching Science and Technology with Museums* http://www.museoscienza.it/smec.

Whitty, Helen 1999, "Making a school excursion a learning experience 1: a work in progress," presented at Musing on Learning seminar, Australian Museum.

Comments

Case Rijsdijk: What ratio of static/interactive exhibits do you feel would be best?

Nick Lomb: Science centers by definition only have interactive exhibits. The Powerhouse Museum and the Sydney Observatory consider it of extreme importance to show real objects – items that people would not have a chance to see or experience outside a museum. We have at least 50–50 real exhibits to hands-on exhibits.

Julieta Fierro: How are lists of "most common questions" in the planetarium compiled?

Martin George: These are based on experience in the planetarium during the question–answer sessions. The list is updated every year or two.

Grant Nicholson: We all noticed how odd Sydney is in not having a permanent planetarium. What is being done to address this situation?

Nick Lomb: The Powerhouse Museum has developed a proposal for a large permanent planetarium on the museum site. It is due to be considered by the government.

19

Science education for the new century – a European perspective

Claus Madsen

European Southern Observatory, Garching D-85748, Germany

Abstract: The paper briefly discusses the surveys about public interest in science and the ways to stimulate interest in science among young people through improvements in the formal science teaching system. It emphasizes the need to develop programs of sufficient size to achieve a long-term impact and obtain the necessary changes. It describes the strategy and individual activities that ESO has undertaken in the field of science education and provides an outlook to the future EIROforum European Science Teachers' Initiative.

19.1   Introduction

Scientists occasionally lament the low public interest in science and the falling level of scientific literacy in the public. However, public surveys continually demonstrate that public interest is high, and although we might not be happy with the current level of scientific literacy, it is, if anything, on the rise (albeit marginally).

In the classical study by Durant and Evans (1989), more than 80 per cent of the respondents professed (a high or moderate) interest in science. A survey in Norway by Eide and Ottosen (1994) found that "as many as 37 per cent of the readers regularly read science articles presented in newspapers." Other researchers using various indicators report significantly higher self-reported interest in science, although there is a strong subject dependency.

Recently, Eurobarometer (European Commission 2001), the survey carried out on behalf of the European Commission in the 15 member-states of the European Union (later augmented by a similar survey in those countries that since joined the Union), looked at attitudes towards and knowledge about science.

Though the aggregate numbers mask very large differences between the individual member-states, 45.3 per cent of all respondents declared interest in science, ahead of politics (41.3 per cent) and economics and finance (37.9 per cent), but below sports (54.3 per cent) and culture (56.9 per cent). However, when asked if respondents feel well informed about a particular area, only 33.4 per cent agree when it comes to science, while the number for sports is 57.0 per cent (i.e., above the number of people who actually declare interest in sports). Of interest – and concern – here, is, of course, the difference between the number of people who claim to be interested in science and those who feel informed. It should be noted that this deficit was already recognized in the study by Durant *et al.* It may be described as the "great paradox," again seen by many scientists as "a problem," though it surely ought to be seen as an opportunity. What it does show, however, is that, for science, public interest does not lead to public engagement. Can it be a sign that current ways of communicating science to the public are not good enough? That we are facing a problem of delivery, not one of lack of interest? In any case, the current situation is clearly unsatisfactory on two accounts. Firstly, a lack of understanding of science poses a problem when citizens are required to take informed decisions on socioscientific issues, thereby endangering the possibilities to reach consensus on such issues, as required in a democratic society. Secondly, the current recruitment level of young people into scientific professions is insufficient to fill the long-term needs of society, a problem that is exacerbated by the demographic development that can be observed in most Western societies.

19.2   Activities to stimulate public interest in science

Recognizing this problem, scientists and science communicators have over the last decade or two become increasingly open to innovative ways of public communication, and terms such as "public understanding of science" or "public awareness of science" have been added to the professional vocabulary.

The proliferation of science centers with interactive "hands-on" displays, and planetariums with their unique instruments, are manifestations of this movement to stimulate public interest in science. For all their merits and the numerous, positive experiences reported by individual visitors, however, we have yet to see a measurable effect in surveys investigating the general level of scientific literacy in the populations of Europe.

19.3   Science teaching in the formal education system

In spite of the relatively high interest in science, why is it that many young people feel disenchanted with science teaching at school? Sjoberg (2003) claims that this has to be understood through social, political and cultural changes, in particular

- changes in the life and world of the young (youth culture);
- changes in science itself, the image and perceptions of science, and science and its relationship to society.

Sjoberg specifically points to a disalignment between the primary elements of contemporary youth culture and features of science teaching at school.

Yet when we talk to some of those young people who opted for scientific careers, many refer to the positive experience at school, saying that "I had a very good science teacher," testifying to the key role of the individual teacher. Indeed, we see many brilliant science teachers, but they are often isolated, rather like bright but dispersed stars on the dark night sky.

There are many obstacles faced by the European corps of science teachers, including the fact that, with education remaining under the authority of the member-states (of the European Union), there is no single European education system. Those member-states with a federal structure may even have several independent systems within their national borders. Germany, for example, has 16 individual education systems.

Furthermore, the rigidity of the education systems poses a major problem for innovative teachers, since the daily pressure of fulfilling the requirements of the established curriculum does not leave much room for individual solutions, or for teaching experiments.

The possibility to change matters is hampered by the relatively low social standing of teachers, a phenomenon that has developed over time and which both limits the political "power" of the teaching community and, in fact, often discourages any attempt on the part of the teachers themselves to remedy the situation.

Finally, for science teaching, there is no mechanism or structure to ensure regular contact between teachers and scientists. With science progressing at a breathtaking pace, the gap between science as it is conducted today and science as it is taught at school is bound to widen, to the detriment of engaged, interest-catching science teaching delivered to pupils at school.

19.4   European Southern Observatory's educational activities

Recognizing the severity of the educational problems and their long-term consequences for science itself, European Southern Observatory (ESO) turned its attention to science education in the early 1990s. ESO's terms of references, as laid down in the Convention (1962), do not provide for comprehensive activities in the field of education, even if in the wake of its regular activities, a wealth of material and useful experience has accumulated that could be put to good use in the schools. Fortunately, through the good will and funding provided mostly by the European Commission, it became possible for ESO to build up educational activities in a systematic fashion. The activities have followed two strands:

- activities for young people;
- activities for teachers.

Realizing the magnitude of the problems, however, it was clear that ESO could not by itself hope to achieve any significant impact on a continental scale. Hence ESO adopted a strategic approach involving the development of pilot programs that could be tested in "real-life" and then, in collaboration with partners, move to larger-scale activities that may achieve the full impact for which we had hoped.

ESO's main partners in these activities have been:

- the European Union, represented by the Directorate General for Research of the European Commission;
- the European Association for Astronomy Education (EAAE);
- the EIROforum (the collaboration between the seven European Intergovernmental Research Organizations that operate major science infrastructures).[1]

The European Association for Astronomy Education (EAAE) is an association of mostly physics teachers from all over Europe. EAAE works actively to stimulate the interest in astronomy among young people. Its members also interact with educational authorities in order to strengthen the physics teaching in their respective countries. The joint effort of scientists and educators has led to the introduction of astronomy as a formal subject in many secondary schools. Moreover, the EAAE constitutes an efficient network inside the education system for ESO as well as for other science organizations in Europe.

Together with one or more of the partners mentioned above, ESO has carried out a series of educational activities in the framework of the European Science Week, an initiative by the European Commission. The activities were primarily aimed at secondary-schoolteachers and pupils. Further, in the year 2001, ESO established a dedicated educational office (http://www.eso.org/outreach/eduoff/) to facilitate contacts with the teaching community

---

[1] CERN (European Organization for Nuclear Research), EFDA (European Fusion Development Agreement), EMBL (European Molecular Biology Laboratory), ESA (European Space Agency), ESO (European Southern Observatory), ESRF (European Synchrotron Radiation Facility) and ILL (Institut Laue-Langevin [neutrons for science]).

as well as conducting seminars for physics teachers, START 2002 (Bacher and West, 2002), and producing a series of teaching materials in collaboration with the European Space Agency (ESA).

Since 1993, ESO has carried out the following programs:

- "The Future Astronomers of Europe" – an essay contest with the title "A night with the VLT" (1993). This contest was organized five years ahead of the Very Large Telescope (VLT) "First Light," to raise awareness of this telescope among young people, some of whom could be expected to graduate from university around the time the VLT would become operational. The national first-prize winners, 18 astronomy-interested young people from as many countries, were invited to perform real observations at the La Silla observatory (West, 1994).
- "Astronomy – Science, Technology, Culture" – an international conference for about 100 secondary-schoolteachers of astronomy (1994), leading to the subsequent forma- tion of the European Association for Astronomy Education (EAAE); see the EAAE website and the "Declaration on Teaching of Astronomy in Europe's Schools" at URL: http://www.eaae-astro.org/.
- "Europe towards the Stars" – a multi-option contest of essays, practical astronomical projects, etc. (1995), carried out in collaboration with the EAAE. About 45 students and their teachers were invited to the ESO Headquarters in Garching (Germany), from where they made astronomical observations with two telescopes at La Silla via the satellite link.
- "Astronomy OnLine" – the first major educational program for astronomy, entirely based on the Internet (1996), in collaboration with EAAE; see West and Madsen (1997). In spite of the fact that the Internet was still a relatively new feature at the time, there were about 5,000 registered participants in 720 groups from 39 countries.
- "Sea and Space" – an Internet-based program with a closing event at EXPO 1998 in Lisbon (1998). Teams of pupils and teachers were invited to produce "a newspaper" describing astronomical aspects of the theme "Sea and Space." The program involved a collaboration with ESA and EAAE; see Madsen and West (1998).
- "Physics on Stage" – a European program (2000) with CERN and ESA as main partners (as well as participation by the EAAE and the European Physical Society (EPS)), aimed at improving and stimulating physics teaching. The final event, the "Physics on Stage Festival," took place in November 2000 at CERN in Geneva (see below).
- "Life in the Universe" (2001), with EAAE, ESA, CERN and ESRF. A contest-based activity for teams of pupils and teachers dealing with the subject of extra-terrestrial life (see below).
- "Physics on Stage II – Focus on Teachers" (2002), with EIROforum and EAAE.
- "Catch a Star" (2002). An Internet-based contest carried out with EAAE (see below).
- "Sci-tech/Couldn't be without it!" (2002). A joint EIROforum activity aimed at primarily young people. The program queried participants about their "favorite" technological items, such as mobile telephones, CD-players, etc., and traced the fundamental research that had been necessary to develop such items. The program was split into a broadly accessible Internet activity, completed by live web-casts, and a professionally conducted survey among the European public.
- "Mercury Transit" (2003), information, including a hotline, about the transit of Mercury on May 7, 2003; see Boffin and West (2003).

- "Physics on Stage III – Physics and Life" (2003), with EIROforum, EAAE and EPS. The subtitle indicates the gradual expansion to cover a broader range of the natural sciences.
- "Catch a Star" (2003), with EAAE.
- "Venus Transit 2004" – a major outreach program with EAAE, IMCCE (France), and the Ondrejov Observatory (the Czech Republic). The activity involved thousands of people across Europe and even beyond the continent in observing the transit of Venus on June 8, 2004.
- "Science on Stage" – with EIROforum, EAAE and EPS.

As mentioned, ESO's activities are directed towards secondary-school pupils and teachers. By way of example, the two different kinds of activities are described in more detail.

19.5   For young people: "Life in the Universe"

"Life in the Universe" was a predominantly Internet-based activity inviting teams of pupils and teachers in 23 countries to make contributions dealing with this subject. Aside from being one of the most basic "great questions" of humanity and having a strong public appeal, it also provided an opportunity for exposing the public to a young and rapidly developing research area. The subject is clearly multidisciplinary and contains strong international aspects. It is a field in which the natural and humanistic sciences come together, it illustrates the interplay between science and technology, and touches on the issue of science versus science fiction. Last, but not least, it has a great potential for media interest – allowing information about the program to spread wide and fast.

The choice of this subject illustrates the deliberate attempt to exploit themes of great interest to young people to attract them to science. As always for these activities, the program was supported by carefully planned public relations and media activities, including posters, flyers, press releases, and the provision of information material (including video-trailers in broadcast standard) to the national committees, etc.

The program itself peaked in a week-long final event, allowing all the national winners to meet with a series of internationally renowned scientific experts in an elaborate program aimed at maximizing the exchange and interaction among the young participants and the scientists.

19.6   For young people: "Catch a Star"

Whereas most of ESO's activities for young people are demanding and therefore primarily tend to attract some of the brightest students, "Catch a Star" allows participation by a broad range of pupils. The requirements are lower and the chances of "success" do not depend on the quality of the entry, provided it passes a reasonable, pre-defined threshold. Nonetheless, prizes are as attractive as for the other programs, for example, involving visits to ESO's observatories in Chile.

19.7   For teachers: "Physics on Stage"

The aims of "Physics on Stage" are to facilitate exchange of best practice among teachers; to provide a forum for in-depth discussion of the problems with which science teachers are faced; to facilitate direct contact among active scientists, working at Europe's foremost science institutes, and teachers; and to provide a "feed-back" mechanism to ensure that the findings and issues identified in the discussion fora are fed into the political process providing the frame for education policies. Finally, raising the societal status of science teachers is an explicit goal of the program, providing visible recognition of their importance for the future development of society.

The "Physics on Stage" program is carried by National Steering Committees in more than 20 European countries. The main mechanism for identifying innovative and engaged science teachers is through national competitions, by which – typically at a national event – teachers are invited to present their ideas that can range from simple exercises or pieces of equipment to elaborate theatrical presentations. In certain cases, the program can even fund the development of specific new activities.

The teachers with most promising contributions are invited to attend the "Physics on Stage" festival, where about 500 teachers from all the participating countries meet. At the festival participants can present their work at the "Science Teaching Fair" as well as through formal on-stage presentations. They can also take part in a series of workshops dealing with a diverse set of topics (such as physics teaching in secondary schools, the history of science as part of the curriculum, women and physics, mathematics and physics, working conditions for teachers, how European organizations can contribute to raising the interest in science among young people, etc.).

Through lectures by active scientists, occasionally including Nobel laureates, teachers are exposed to the world of modern science.

Last but not least, the festival offers a prestigious prize for the best science teaching initiative, bestowed with 10,000 euros. The first prize, in 2002, was handed over by the European Commissioner for Research, Philippe Busquin.

The festival attracts not only teachers. The organizers are also keen to involve administrators from the educational authorities and politicians from all member-states and the European Parliament.

Of utmost importance is the feedback mechanism, ensuring that the results of the festival are made available for the entire science teaching community, administrators, and politicians. For this purpose, the national committees set up post-festival evaluation and information meetings in the participating countries.

Recently, the organizers have decided to broaden the scope of "Physics on Stage" to accomplish a wider range of the natural sciences (notably biology). To mark this change, the program is now changing its name to "Science on Stage." In the context of astronomy education, it should be stressed, however, that this in no way diminishes the importance or visibility of astronomy in the program. Rather it allows the multidisciplinary character of our science to show its full potential.

19.8   The EIROforum European Science Teachers' Initiative

The unqualified success of "Physics on Stage/Science on Stage" has led EIROforum to propose a long-term educational activity, with "Science on Stage" as the core activity, but expanded with a series of additional, complementary activities. These may include:

- summer schools at the EIROforum member organisations;
- teachers-in-residence at the EIROforum member organisations;
- scientists giving talks at schools;
- the development of an Internet-based repository of teaching materials.

The EIROforum Science Teachers' Initiative, which will be open to about 30 countries, is currently being considered for possible co-funding by the European Commission.

19.9   Conclusion

ESO's involvement in science education has been based on a fairly altruistic attitude, driven by a general concern about the science literacy in the public and a realization of the potential for public exploitation of its rich collection of data and general experience. Nonetheless, the increasingly precarious recruitment situation for future scientists provides a strong additional driver for this engagement. In spite of the formal limitations as given by ESO's terms of reference ESO's willingness to open up for education initiatives and to engage with the education community has been welcomed by its member-states.

In addressing the problems of scientific literacy and public interest in science, we recognize the need for large, coordinated programs aimed at different groups in society, for example, professional teachers and young people. Such programs require investments in money and manpower that exceed the capabilities of most science organizations. Therefore, cooperation among many partners and the joint involvement of science and the education community is called for.

Astronomy is particularly well suited for such activities, thanks to its multidisciplinary character and its natural attractiveness for young people.

Through their consistency, the ESO-backed educational activities are increasingly seen as a "permanent" feature in the move to stimulate interest in the natural sciences in general, and in astronomy in particular. At the same time, the catalyst effect of the European Science Week and the "Science and Society" program of the European Commission should be acknowledged for enabling such activities.

References

Bacher, A. and West, R. M. 2002, "First teachers' training course at ESO HQ was a great success," *The ESO Messenger*, **109**, 57–8.

Boffin, H. and West, R. M. 2003, "The May 7 Mercury transit," *The ESO Messenger*, **112**, 57.

Durant, J. and Evans, G. A. 1989, "Understanding of science in Britain and the USA," in *British Social Attitudes: Special International Report*. R. Jowell, S. Witherspoon, and L. Brook (eds.). Aldershot: Gower.

European Commission, 2001, "Europeans, science and technology," Eurobarometer 55.2, the Directorate-General for Press and Communication, Public Opinion Sector.

Madsen, C. and West, R. M. 1998, "Sea and space – a successful educational project in Europe's secondary schools," *The ESO Messenger*, **93**, 44–6.

Sjoberg, S. 2003, "Why don't they love us any more? Perspectives on science in education and in society," presentation given at the Center Albert Borchette, Brussels, in connection with the Science Education Initiative by the European Commission.

West, R. M. 1994, "Future astronomers of Europe," *Sky and Telescope*, **88**(3), 28–32.

West, R. M. and Madsen, C. 1997, "The astronomy on-line project," *The ESO Messenger*, **87**, 51–4.

Comments

Case Rijsdijk: Regarding the "popularity" of science and technology: do you think that people perceive technology as computers, and that this perception influences the percentage of interest?

Robert Hollow: Having had my Australian students involved in the 1996 Astronomy On-line observation of a lunar eclipse, I am impressed by the valuable perspective in real-time communication among students in different languages. While my students could not observe the eclipse because of heavy rain, they realized that Italian schools could, when "splendissimo" appeared on their email. Some students studying French were busy trying to translate messages for the other students. This was a very valuable learning experience. Do you at the ESO try to work with language teachers and educators to facilitate communication?

Rosa M. Ros: It is a good idea to promote more than one official language for teachers. The EAAE summer school has three official languages, and this situation does not introduce any specific problems.

20

Communicating astronomy to the public

Charles Blue

National Radio Astronomy Observatory, Charlottesville, Virginia, USA

Abstract: Increased interest among education and public outreach specialists in coordinating the task of "Communicating Astronomy to the Public" led to a conference of that name in Washington DC, following an earlier conference in Tenerife the previous year. One outcome of the conference was the Washington Charter, which is to help foster such public education through statements of principles of action for funding agencies, professional astronomical societies, universities/laboratories/research-organizations, and individual researchers, respectively. Another outcome was the formation of a Working Group on the subject within the Union-wide Activities Division of the International Astronomical Union. Information about both outcomes can be found at http://www.communicatingastronomy.org.

Editors' Note: This paper was solicited by the editors in March 2004.

Nearly 250 outreach professionals in the astronomical community gathered in Washington, DC, on October 1–3, 2003, to attend the "Conference on Communicating Astronomy to the Public." This three-day conference attracted public information officers, astronomers, educators, and members of the entertainment and news media to explore the gaps in outreach, the current and emerging demands of the public, the needs of the astronomical community, and to work on methods to answer these needs. "Education and Public Outreach," now often abbreviated as "E/PO," was the overall subject.

The conference was organized by the National Radio Astronomy Observatory (NRAO), for which I am a public information officer, and hosted by the US National Research Council. The morning sessions of this conference were based on a series of panel discussions, addressing such topics as astronomy in entertainment, image repositories, best practices, and underdeveloped audiences. More specifically, titles covered included: "Entertainment and Hype in Outreach"; "Astrophysics vs. Astrographics" (or how to represent astronomical data graphically); "Needs of Underserved Outreach Avenues" (what small planetariums, science centers, park rangers, and others need from the astronomical community); "the Connection Between E/PO and Research Astronomy"; and "Successful Practices in Communications." Each afternoon featured various breakout sessions, organized around a series of questions generated during the morning panels.

The results of this conference are still bearing fruit through a number of ongoing activities. The first outcome was the Washington Charter, which stresses the need for astronomy organizations to place sufficient emphasis on outreach, and provide the means to make a real contribution to this activity. The charter is being disseminated among many organizations, with the recommendation that it be adopted officially. Additionally, five full members of the American Astronomical Society (AAS) sent a formal request to the AAS Executive Committee that they adopt this document as an official position of the society.

The Charter is available at:

(http://www.communicatingastronomy.org/washington_charter/index.html) and reads as follows:

THE WASHINGTON CHARTER FOR
COMMUNICATING ASTRONOMY WITH THE PUBLIC
The Statement of the "Communicating Astronomy to the Public"
Conference (October 1–3, 2003)

CHARGE: As our world grows ever more complex and the pace of scientific discovery and technological change quickens, the global community of professional astronomers needs to communicate more effectively with the public. Astronomy enriches our culture, nourishes a scientific outlook in society, and addresses important questions about humanity's place in the universe. It contributes to areas of immediate practicality, including industry, medicine, and security, and it introduces young people to quantitative reasoning and attracts them to scientific and technical careers. Sharing what we learn about the universe is an investment in our fellow citizens, our institutions, and our future. Individuals and organizations that conduct astronomical research – especially those receiving public funding for this research – have a compelling obligation to communicate their results and efforts with the public for the benefit of all.

PRINCIPLES OF ACTION:
Funding Agencies Should:

- Encourage and support public outreach and communication in projects and grant programs;
- Develop infrastructure and linkages to assist with the organization and dissemination of outreach results;
- Emphasize the importance of such efforts to project and research managers;
- Recognize public outreach and communication plans and efforts through proposal selection criteria and decisions and annual performance awards; and,
- Encourage international collaboration on public outreach and communication activities.

Professional Astronomical Societies Should:

- Endorse standards for public outreach and communication;
- Assemble best practices, formats, and tools that will aid in effective public outreach and communication;
- Promote professional respect and recognition of public outreach and communication;
- Work to promote professional respect and recognition of public outreach and communication;
- Make public outreach and communication a visible and integral part of the activities and operations of the respective societies; and,
- Encourage greater linkages with successful ongoing efforts of amateur astronomy groups and others.

Universities, Laboratories, Research Organizations, and Other Institutions Should:

- Declare public outreach and communication a clear priority for all departments and personnel;
- Actively recognize public outreach and communication efforts when making decisions on hiring, tenure, compensation, and awards;
- Provide appropriate institutional support (e.g., funding, infrastructure, personnel, training, etc.) to enable and assist with public outreach and communication efforts;
- Collaborate with funding agencies and other support organizations to help ensure that public outreach and communication efforts are efficient and have the greatest possible impact;

- Develop appropriate formal public outreach and communication training for all researchers; and,
- Integrate communication training (e.g., writing, speaking, etc.) into the academic courses of study for the next generation of researchers.

Individual Researchers Should:

- Actively participate – directly or indirectly – in communicating the results and benefits of astronomical research directly to the public;
- Convey the importance of public outreach and communication to all team members; and,
- Instill this sense of responsibility in the next generation of researchers.

Revised June 2005

Other outcomes include advancing Public Outreach as an official Working Group of the International Astronomical Union (IAU), within Division XII on Union-wide Activities. Also, a working group from the conference is now engaged in developing a shared-resource website with images, outreach materials, and information for the astronomical community. This is to serve as a resource for astronomers and those engaged in outreach.

Details and updates from this conference are located at www.nrao.edu/ccap. (*Editors' Note*: It was also noted at www.astronomyeducation.org.)

Editors' Note: An earlier conference on "Communicating Astronomy" was organized by Terry Mahoney of the Astronomical Institute of the Canary Islands, and held there in Tenerife. See http://www.iac.es/proyect/commast/; a CD-ROM with the proceedings is also available.

Yet another conference on the subject was held in 2005, sponsored by the European Southern Observatory, the European Space Agency, and the International Astronomical Union. See http://www.communicatingastronomy.org/cap2005. An IAU Working Group on "Communicating Astronomy to the Public" has been formed, with website http://www. communicatingastronomy.org.

The Communicating Astronomy with the Public 2005 conference

The Communicating Astronomy with the Public 2005 conference was held in June at the headquarters of the European Southern Observatory (ESO) in Garching/Munich, Germany. The conference was highly successful and covered a range of topics from Case Studies to Audiovisual techniques. The conference was web-cast to the world and all the talks and web casts are found through the conference web page: http://www.communicatingastronomy. org/cap2005.

Apart from plenary talks and discussions a number of hands-on sessions where issues such as how to produce "PR images" from raw data, how to write better texts for the public or even how to make a movie DVD.

Suggested updates to the Washington Charter were discussed and a revised Charter was agreed upon. The proceedings from the conference will be published by ESO during Autumn 2005.

The conference was sponsored by the European Southern Observatory, the European Space Agency and the International Astronomical Union.

Poster highlights

The integral role of planetariums in engaging the public on issues in astronomy is addressed in **On the role of planetariums** by **Anthony P. Fairall** from Cape Town, South Africa.

Probably in excess of 100 million persons, mainly youngsters, pass through the world's planetariums each year, by far the largest, and arguably the most influential, conveyance of astronomy to the general public. However, while some planetariums have close ties to the research world, and even to IAU Commission 46, they are by far the exception and not the rule. In general, the planetarium community operates independently of the stakeholders represented at this conference.

While much of planetarium activity shares a common mission with the IAU Commission on Astronomy Education and Development, there are significant deviations: since the main market driving the planetarium world is clearly throughput, some planetariums emphasize entertainment and novelty more than the teaching of astronomy. There is also an unfortunate tendency in smaller planetariums, where lecturers are weak on science, to overemphasize star lore and constellations. In the author's opinion, the gap between the teaching of astronomy, as seen from the research world of the IAU, and teaching of astronomy, as seen from the planetarium world, badly needs closing.

Members of the IAU Commission on Education and Development might also be made aware of changes in the planetarium world. While the traditional star projector (providing a naked-eye view of the night sky) is far from dying, many new facilities have instead opted to go fully digital, whereby any image (naked-eye view of the night sky included) can be reproduced on the hemispherical dome. The trend is towards "immersive experiences." In the wrong hands, presentations can stray far from astronomy. In the right hands – such as at the Hayden Planetarium in the American Museum of Natural History, New York City – the technology can be used to carry the audience through the galactic neighborhood and beyond. The IAU needs to form a partnership with the planetarium community to find a common path towards the dissemination of astronomical knowledge.

Further issues surrounding planetariums are explored by **Klim I. Churyumov** in a poster entitled **Kyiv Planetarium and astronomical education in Ukraine**.

The main task of the Kyiv Republican Planetarium, aside from lecturing on astronomy and space physics for the general public, is to provide support for teaching astronomy in high schools. In the planetarium, educational astronomy programs are designed so that they are closely connected with the teachers of Kyiv and are intended to introduce certain additions to the traditional curriculum. They allow a better understanding and a deeper study of numerous astronomical phenomena and physical mechanisms of cosmic processes. In the planetarium's educational programs, we use up-to-date scientific information about new discoveries in astronomy obtained with the world's largest telescopes, including the Hubble Space Telescope (HST).

Planetariums are but one example of situations where students might come in contact with astronomy for the first time. The importance of this encounter is explored in **First contact with astronomy for a large number of pupils** by Spain's **Rosa M. Ros.**

The Spanish Royal Society of Physics (RSEF) co-operates with several European institutions to promote physics and astronomy in schools through the project "Física en Acción." This project started in 2000, integrated with the project "Physics on Stage," created by CERN, ESA, and the ESO. "Física en Acción" is a Spanish competition bringing together a group of teachers in a common endeavor:

- showing "physics demonstrations" to general audiences; and
- engaging in pedagogical presentations to introduce science into the classroom.

The national final event of this competition takes place annually in a science museum during one weekend (entrance is free). The Science Fair is especially well received by visitors, who can ask the demonstrators-teachers questions. Younger visitors enjoy experimenting for themselves. After the first year, the RSEF introduced special prizes to encourage schools to participate in astronomical categories. The "Centro de Astrobiologia de Madrid" gave a cash prize and a visit to their headquarters to the winners. The "Instituto Astrofísico de Canárias" offered a prize of a trip to its observatories. In summary, the astronomical elements of "Física en Acción" stimulate the teachers' and students' interest in international activities and have been the first contact with astronomy for a large number of pupils.

This initiative had a warm welcome from teachers, especially because the fair promoted face-to-face contact between them, whether they were active participants in the project or only visitors. It was also welcomed by students, more of whom visit every year with their teachers.

A poster discussing plans to harness the educational power of the 2004 Venus transit entitled **The 2004 Venus transit: a European educational project** is presented here by **Jean-Eudes Arlot.**

On June 8, 2004, the planet Venus passed in front of the disk of the sun. This rare event (no one alive today had seen such a transit) reminds us of the story of the measurement of the size and scale of the solar system.

We used this event to organize a worldwide network of high schools, individuals, and scientific centers to carry out an accurate timing of the event of June 8. Moreover, we helped pupils, students, and the general public to understand a scientific procedure needing an international collaboration and to be aware of the powerful tool that is a transit, and that is now being used for detection of extra-solar planets.

We organized a centralized computation of the Astronomical Unit through the Internet, thanks to individual timings of the event, and provided notes and educational material to participants, in order to encourage their interest in science, and to promote safe observation of the sun.

The transit was widely seen, in generally favorable weather, and the results are available through a link on the IAU Commission on Education and Development's special Website at http://www.transitofvenus.info, or directly at the European Southern Observatory's special Website at http://www.vt-2004.org/auresults/. See also http://www.imcce.fr/vt2004/ for discussions in English and in French.

A contribution from **Mary Kay Hemenway** *et al.* concerns **Student pre-post-visit materials at McDonald Observatory**.

The "Student Field Experience" program at McDonald Observatory (Fort Davis, Texas) is designed to allow students to connect their visit to a research observatory to school science and everyday life. To enrich their experience, materials have been developed for the teacher to use in the home classroom both before and after their visit. Evaluation has shown that pre-visit materials increase students' curiosity and prior knowledge so that they are ready to be immersed within the observatory environment. The pre-visit package includes a map, observatory rules, student behavior expectations, and suggested activities. For example, one activity has them contrast their hometown community with the observatory as a community. They also reflect on the differences between telescope, dome, and observatory.

During their visit, they tour a research telescope and explore the exhibit hall ("Decoding Starlight" on spectroscopy). Options include participation in an evening star party, and, for smaller groups, a hands-on activity chosen from a menu of topics. Pre-post-visit suggestions are provided to the teacher for items in this menu. Student Exhibit Guides (for grades K–3, 3–6, and 6–12) emphasize topics related to science standards. A teacher key to the student guides offers in-depth explanations of the concepts and suggestions for further activities.

Post-visit suggestions refer back to the pre-visit challenges. In order to better serve students with limited English proficiency, some of the materials have been translated into Spanish. The post-visit materials encourage student reflection on their experiences. An Educator Advisory Board of current and former K–12 teachers contributes to the formative evaluation that allows changes to be made over time. A summative evaluation by an outside evaluator is in process.

Support for HST-ED90234-01 was provided by NASA through a grant from the Space Telescope Science Institute, which is operated by the Association of Universities in Astronomy, Incorporated, under NASA contract NAS5-26555.

A representative from Mexico, **Hector Bravo-Alfaro**, explores efforts at his university to embark on **A project for a center for astronomy popularization**.

By 2000 the facilities of a rustic observatory, named "La Azotea" (the Roof), were formally assigned to the Astronomy Department of Universidad de Guanajuato, in Mexico (about 400 km north-west of Mexico City), where eight professional astronomers do research and undergraduate-level teaching. This center, built in the late 1970s on the top of the main building of the university, disposed of a 16-cm refractor telescope that had not been in use for about 12 years because of electronics problems. Only a portable 8-inch Schmidt–Cassegrain exists as equipment.

The whole complex consists of a telescope dome, a classroom with capacity for some 50 people, an office, and a photography darkroom. In general, maintenance was non-existent, and no recovery program had been considered at the time we took charge of the site. Since then, we have developed a program of free astronomical observations for non-specialized audiences, in particular devoted to students. With very limited funds we got a second portable 8-inch Schmidt–Cassegrain, binoculars, solar filters to be used with the portable telescopes, slide collections, and a modern PC, but it remains to make big repairs to the building itself, carry out electronic tuning of the refractor telescope (the optics are perfect), and acquire the necessary furniture to take advantage of the hall as conference room.

To date, we have designed a project to develop a Center for the Popularization of Science at this location with the aim of setting up the following activities:

- permanent program of astronomical observations for students;
- regular conference series in every science domain;
- summer schools in astronomy for elementary and high-school teachers;
- foundation of an amateur society of astronomy.

The Astronomy Department of Universidad de Guanajuato is currently in charge of the astronomy branch of the undergraduate program of physics, with a graduate program in astronomy starting in summer 2004. We are planning that students in the first levels of these programs do part of their observational training with the help of these facilities. Students in advanced levels of the undergraduate program will participate in the activities of the center as part of their social duties.

Further exploration of issues concerning the popularization of astronomy with the general public is found in **Raising awareness in society on space at Bosscha Observatory** (Indonesia) by **Suhardja D. Wiramihardja**.

Besides the main tasks in research and teaching astronomy, we at the Bosscha Observatory are obliged to bring astronomy to society by giving public lectures to people in an organized program. Because of the varied education backgrounds of those visiting Bosscha Observatory, observing assistants – advanced-year students – are involved in the activities as well. The program mainly consists of popular astronomical lectures using slides or a computer display with question–answer session and an inspection of the main Zeiss 60-cm double-refractor telescope. In the dry season in the period of April–September, a session to observe stars, planets, and the moon with smaller telescopes is offered to the visitors, accommodating their curiosity about how stars, planets, and the moon may look like through a telescope.

This is one of the ways to apply a scientific approach to understand nature. We hope that in the long run this program can raise awareness in the community about space and will enhance the appreciation of science, especially astronomy, which is usually given low priority in developing countries like Indonesia.

Two posters are contributed here by **Jun-ichi Watanabe**, the first concerning **Star Week – a successful campaign in Japan**.

There are several hundreds of astronomical facilities for the general public in Japan, including planetariums, museums, and public observatories. In order to connect these facilities, and to make good collaborations for educational purposes, the "Star Week" was originally proposed by the author in 1995, and has continued up to the present.

Star Week was originally set to take place during August 1–7, when it is usually expected that most of Japan should have good weather after the rainy season. In 1995, more than 100 facilities participated in this campaign, and about 200 astronomical events such as star parties were coordinated. These numbers are increasing every year, and 219 facilities performed 484 events in 2002. The activities of Star Week are among the most successful campaigns for the public to experience stars firsthand.

Next we have a second contribution from Watanabe, entitled **JAASC Cooperation League for education and public outreach**.

The Japanese Astronomy, Aeronautical Science, Space Science (JAASC) Cooperation League was established in 2000 among the related institutes for education and public outreach. The participating institutes are the National Astronomical Observatory of Japan, the Institute of Space and Aeronautical Science, National Aerospace Laboratory of Japan, Young Astronomers Club, Japan Science & Technology Corporation, and Japan Space Forum. These institutes started several joint efforts, such as creating websites for beginners in the general public, or providing educational materials for junior high school. This is a challenging trial for Japanese institutes to cooperate beyond the barrier of the "bureaucracies."

An exemplary freeware program that allows the public to experience "astronomy in 3D" is explained in **Cosmo.Lab: stereographic viewing of astronomical data** by **Hans R. De Ruiter** *et al.* from Bologna, Italy.

The creation of large data sets with the new generation of large telescopes will require new display tools. Since practically all such data are multidimensional, the need for stereographic viewing of astronomical data will steadily grow. However, 3D display may also quickly become a fascinating tool in education and outreach, the more so because modern computer technology already permits us to implement such viewing on personal computers at very low cost (a few hundred US dollars). The almost infinite possibilities of 3D viewing motivated us to develop a software package that allows 3D viewing and manipulation of very large data sets. These data sets may be in the form of catalogs in simple ASCII format, or in other formats used in astronomy (e.g., FITS, Tipsy): both gridded data and simple lists of objects can be handled.

The software package, "Cosmo.Lab," is easy to use, fast, and user-friendly. The project, a collaborative effort of CINECA (an inter-university institute for computing, based in Bologna, Italy), the Institute of Radio Astronomy (Bologna), the CNR Institute of Cosmic Physics in Milan, the astronomical observatory of Catania, and ASTRON (the Netherlands), is funded by the European Union and is at present in an advanced stage of completion. It can run under different operating systems (Linux, MS Windows). It is freeware and can be freely downloaded from the Cineca website (http://cosmolab.cineca.it). Upon request, a CD-ROM will also be available. It is possible to take "guided tours" through the data, for example through the HIPPARCOS Catalogue or a redshift catalog, because Cosmo.Lab permits recording of the various steps one wants to make; thus one can construct "movies." Some fixed guided tours will be part of the package itself, which means that Cosmo.Lab should be easy to use also for those educators who are unfamiliar with computer programming. At the Cosmo.Lab website, more information on the project can be found; in addition to the software, one can also download the user guide (in pdf format).

The use of the Internet to allow amateurs to remain connected and informed is recounted in **The most popular astronomical web server in China** by **Chenzou Cui**.

Affected by the consistent lows in the IT market, free homepage space has become less and less available in China. It is more difficult to construct webpages for amateur astronomers who do not have the ability to pay for commercial space. In May 2003, with the support of the Chinese National Astronomical Observatory and the Large Sky Area Multi-Object Fiber Spectroscopic Project, we set up a special web server (amateur.lamost.org) to provide free, huge, stable, and non-advertised web space to Chinese amateur astronomers and non-professional organizations. After only one year, there were more than 10,000 visitors from nearly 40 countries, and the amount of data downloaded exceeds 4 GB/day. The server has become the most popular amateur astronomy web server in China. It stores the most abundant Chinese amateur astronomical resources. Because of the extreme success, our service has been drawing tremendous attention from related institutions. Recently, the Chinese National Natural Science Foundation showed great interest in supporting the service. In our paper, we examine the emergence of the idea, describe how we constructed the server and its present utilization, and introduce our future plans.

A new television program in Cuba with the goal of increasing scientific literacy in the general public is introduced in **Teaching astronomy through TV in Cuba** by **Oscar Alvarez**.

In an effort to elevate the scientific culture of the Cuban people and to serve as support to the National Education System, a new channel of the National Television Network has been created dedicated exclusively to public education. One of the first courses was "Foundations of Modern Science," in which several conferences on topics such as astronomy were addressed. In this presentation, we show the actions that we are taking in the country as a contribution to the teaching and public understanding of science in general, and astronomy in particular.

The last poster in this section is from Thailand's **Yupa Vanichai**, entitled **Our attempts in astronomy**.

During the last decade of the twentieth century, astronomical articles in Thai scientific magazines remained out of date. Topics such as interacting galaxies, black holes, and other celestial objects beyond the solar system were hardly found in the media. However, a pocketbook about deep space was written by a lecturer, and a website of astronomy for the Thai people was planned through the cooperation of two computer programmers. An observatory with a 600-mm reflector was built by a Thai engineer. A 150-mm Dobsonian reflector, made in Thailand, is sold at a low price. Future optical programs are also being planned.

Part IX

The education programs of the International Astronomical Union

Introduction

The International Astronomical Union (IAU) was founded in 1922 to "promote and safeguard astronomy... and to develop it through international co-operation." There are currently (2005) 9,014 individual members in 87 countries. The IAU is funded through the adhering countries, and is administratively "lean." The total staff consists of a secretary and an assistant. The officers serve voluntarily, usually with support from their academic institutions. Almost all of the funds supplied from the dues are used for the development of astronomy.

One of the 40 IAU "commissions," or interest groups, is Commission 46, formerly called "The Teaching of Astronomy," and more recently, at the 2000 General Assembly, renamed "Astronomy Education and Development." It is the only commission that deals exclusively with astronomy education; a previous Commission 38 ("Exchange of Astronomers"), which allocated travel grants to astronomers who need them, and a "Working Group on the World-wide Development of Astronomy," have been absorbed by Commission 46. The 40 commissions, and the many working groups, were recently organized into 11 scientific divisions. Commission 46 is part of a 12th division, "Union-wide Activities," which deals with issues of concern to all IAU members.

The commission's mandate is "to further the development and improvement of astronomy education at all levels, throughout the world." In general, the commission works with other scientific and educational organizations to promote astronomy education and development; through the national liaisons to the commission, it promotes astronomy education in the countries that adhere to the IAU; and it encourages all programs and projects that can help to fulfil its mandate.

The commission holds business sessions at each IAU General Assembly. Within the new format of the IAU General Assemblies, the commission organizes or co-sponsors major sessions on education-related topics, such as the session on which this book is based. The commission has also organized two major conferences on astronomy education – in the USA in 1988, and in the UK in 1996. For three decades, the commission has sponsored one-day workshops for local schoolteachers, as part of every IAU General Assembly, and as part of several IAU regional meetings. Immediately after the conference that is described in this book, a very successful teachers' workshop was held in Sydney, organized by Nicholas Lomb, Sydney Observatory. Until recently, Commission 46 was concerned primarily with tertiary (university-level) education and beyond, but several of its activities have an impact on school-level and public education. The Sydney conference may spur a greater interest in these levels.

One of us (JMP) is the current president of Commission 46; the other of us (JRP) is a former president. In Chapter 21, Syuzo Isobe, who was president of Commission 46 from 2000 to 2003, describes its current structure and work.

You can find out more about IAU Commission 46 by visiting the commission's website at: http://physics.open.ac.uk/IAU46/; which is equivalent to: http://www.astronomyeducation.org.

21

A short overview of astronomical education carried out by the IAU

Syuzo Isobe

National Astronomical Observatory, 2–21–1, Osawa, Mitaka, Tokyo, Japan

Abstract: Astronomy education is one of the important vehicles to make people aware of the role of science in daily life and in environmental issues. Therefore, in a good school curriculum, astronomy should be included. In the process, however, some governments try to introduce a system that succeeds in other countries, without considering cultural and national differences. Unfortunately there is no program group of IAU Commission 46 (on Education and Development) that deals with astronomy education in schools. It is hoped that such a program group will be set up after the conference on which this book is based.

21.1 An overview of IAU Commission 46: "Astronomy Education and Development"

The International Astronomical Union (IAU) is a union of professional astronomers who produce a large number of new astronomical results and make the frontiers of astronomy expand. However, the union cannot stand by itself but needs much support from governments as well as from the public. Therefore, the IAU established Commission 46 through the vigorous efforts of E. Schatzman, and others. In 1967 it was originally called "Teaching of Astronomy" and was renamed in 2000 "Astronomy Education and Development," with a much wider scope of activities, in order to cover from the level of beginners to that of doctoral and post-doctoral students. Table 21.1 gives a list of its presidents; they are well distributed internationally.

There are currently nine "program groups," as shown in Table 21.2. Because of limited budgetary resources, it is difficult for the IAU to cover all the possible fields to promote our Commission's goals. The "National Liaison" Program Group communicates closely with the Organizing (Executive) Committee members as well as with the regular members by producing national reports from individual countries every three years. All of the committee activities and some related activities are communicated to the members, and to other people interested in astronomy education, mainly through the Internet and electronic mail (but in some special cases by printed version) by the "Newsletter" Program Group.

The "Public Information on Eclipses and Transits" Program Group holds public lectures and provides information related to each individual total or annular solar eclipse, including how to look at the eclipse safely.

For a country, especially a developing country, intending to develop astronomy and astronomical education it is important to have at least one – but hopefully several – science (mainly physical science) professors in that country who have a strong motivation to develop it. The "Worldwide Development of Astronomy" Program Group works to find such professors through different channels of the program group's members. Once this program

Table 21.1. *The list of past presidents of Commission 46*

1967–1970	E. Schatzman	France
1970–1973	E. A. Müller	Switzerland
1973–1976	D. McNally	UK
1976–1979	E. Kononovich	USSR
1979–1982	D. G. Wentzel	USA
1982–1985	L. Houziaux	Belgium
1985–1988	C. Iwaniszewska	Poland
1988–1991	Aa. Sandqvist	Sweden
1991–1994	L. Gouguenheim	France
1994–1997	J. Percy	Canada
1997–2000	J. Fierro	Mexico
2000–2003	S. Isobe	Japan
2003–2006	J. Pasachoff	USA

Table 21.2. *Program groups of Commission 46*

1.	Worldwide Development of Astronomy
2.	Teaching for Astronomy Development
3.	International Schools for Young Astronomers
4.	Exchange of Astronomers
5.	Collaborative Program
6.	Newsletter
7.	National Liaison
8.	Public Information on Eclipses and Transits
9.	Exchange of Books and Journals

group identifies a specific country that is ready for development, such information is then sent to the "Teaching for Astronomy Development" Program Group which starts to send lecturers to that country and to invite a number of students to an institute in a developed country where high-level astronomical research is carried out. Those students are expected to promote astronomical education in their home countries later.

Typically, developing countries have only one or a few professors, whose astronomical fields may be very limited. To give graduate students and young astronomers a wider perspective, the "International Schools for Young Astronomers" Program Group organizes schools in different regions. A list of past schools is given in Table 21.3. If a young astronomer intends to carry out research at a specific institute in a more developed country, the "Exchange of Astronomers" Program Group partially supports his/her travel costs. Since the IAU budget is limited, the "Collaborative Program" group seeks the possibility to carry out activities in partnership with other organizations. The "Exchange of Books and Journals" Program Group facilitates effective exchanges of this kind.

As shown above, it is an unfortunate situation that our program groups do not directly include astronomy education in schools. I, as the Commission 46 president in July 2003, hope the special session on which this book is based will be a trigger for such a new program group in the future.

Table 21.3. *International Schools for Young Astronomers*

	Time	Site	Weeks	Participants[a]
1	1967 March	UK, Manchester	6.5	12 (12 f, 8 n)
2	1968 June/July	Italy, Arcetri	8.5	10 (10 f, 8 n)
3	1969	India, Hyderabad	8	23 (5 f, 5 n)
4	1970 Oct/Nov	Argentina, Cordoba	8	21 (5 n)
5	1973 Aug	Indonesia, Lembang	4	8 (3 f, 4 n)
6[b]	1974 May	Argentina, San Miguel	4	60 (21 f, 7 n)
7	1975 Sep	Greece, Athens/Thera	4	74 (35 f, 16 n)
8	1977 Nov	Brazil, Rio	4	29
9	1978 Aug	Nigeria, Nsukka	3	28
10	1979 Sep	Spain, Tenerife	2	36 (8 n)
11	1980 Oct	Yugoslavia, Hvar	3	25
12	1981	Egypt, Cairo	3	28
13	1983 June	Indonesia, Lembang	3	21 (5 n, 5 w)
14	1986 Aug	China, Beijing	3	52 (6 n)
15	1986 Sep	Portugal, Espinho	3	30 (19 f, 10 n, 6 w)
16	1986 Aug	Cuba, Havana	2	52 (23 f, 6 n, 7 w)
17	1990 June	Malaysia, Kuala Lumpur and Melaka	2.5	27 (12 f, 8 n, 6 w)
18	1990 Sep	Morocco, Marrakesh	2.5	53 (39 f, 14 n, 7 w)
19	1992 Aug	China, Beijing and Xinglong Obs.	3	30 (17 f, 12 n, 9 w)
20	1994 Jan	India, Pune	3	35 (25 f, 13 n, 11 w)
21	1994 Sep	Egypt, Cairo and Kottamia Obs.	3	41 (12 f, 13 n, 10 w)
22	1995 July	Brazil, Belo Horizonte and Serra Piedade	3	38 (20 f, 11 n, 15 w)
23	1997 July	Iran, Zanjan	3	38 (14 f, 8 n, 12 w)
24	1999 Aug	Romania, Bucharest	3	40 (18 f, 8 n, 22 w)
25	2001 Jan	ChiangMai, Thailand	3	36 (17 f, 9 n, 19 w)
26	2002 Aug	Casleo, Argentine	3	28 (14 f, 8 n, 10 w)
27	2004 Jul	Morocco, Al Akhawayn	3	29 (18 f, 13 n, 9 w)
28	2005 Jul/Aug	Mexico, Puebla	3	

[a] Participants: f = foreign, n = nationalities, w = women.
[b] Three parallel schools.
Notes: 1. ISYA 1 and 2 resembled most summer schools, with local faculty; ISYA 3 onwards emphasized scientifically developing countries, participants from scientifically isolated sites, with about half or more faculty foreign. 2. UNESCO contributed to ISYA 1 through 20. The ISYA 3 through 19 were designated IAU/UNESCO schools. 3. ICSU contributed, explicitly or via IAU, from ISYA 19 onwards.

The ISYA prior to 1992 were organized by Josip Kleczek (Czech Republic) from 1992 until 1997 by Don Wentzel (USA) and Michèle Gerbaldi (France) and since then by Michèle Gerbaldi (France) and Ed Guinan (USA).

21.2   To whom should astronomy be taught?

In recent decades, the development of astronomy is and has been so rapid, and because of interesting new discoveries, many people seem to have an interest in astronomical phenomena. However, in most cases, they are interested in the information but, because of a shortage of astronomical and physical knowledge, their understanding may not be correct.

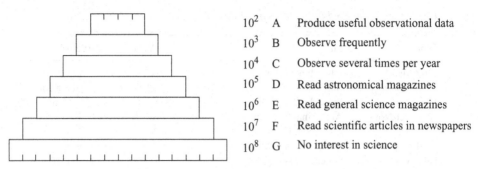

10^2	A	Produce useful observational data
10^3	B	Observe frequently
10^4	C	Observe several times per year
10^5	D	Read astronomical magazines
10^6	E	Read general science magazines
10^7	F	Read scientific articles in newspapers
10^8	G	No interest in science

Fig. 21.1. One example of a structure of education in astronomy. Here people are divided into seven categories, A to G, depending on their level of interest in astronomy. This figure shows a structure in the shape of a pyramid, and is nearly the case in Japan.

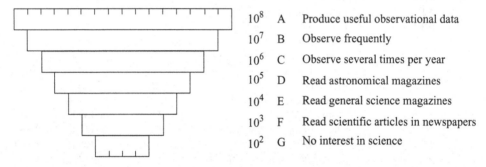

10^8	A	Produce useful observational data
10^7	B	Observe frequently
10^6	C	Observe several times per year
10^5	D	Read astronomical magazines
10^4	E	Read general science magazines
10^3	F	Read scientific articles in newspapers
10^2	G	No interest in science

Fig. 21.2. Same as for Fig. 21.1, but with a different structure. This is not a realistic case.

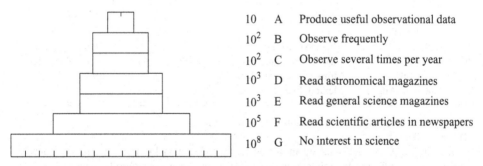

10	A	Produce useful observational data
10^2	B	Observe frequently
10^2	C	Observe several times per year
10^3	D	Read astronomical magazines
10^3	E	Read general science magazines
10^5	F	Read scientific articles in newspapers
10^8	G	No interest in science

Fig. 21.3. Same as for Fig. 21.1, but with a different structure. This structure is frequently seen in developing countries.

Over 10 years ago, I presented a diagram (Fig. 21.1) showing what fraction of people was interested in what level of astronomy for the case of Japan (Isobe, 1991). Our target for the distribution is certainly not the case of Fig. 21.2 and Fig. 21.3 but Fig. 21.4. It is difficult to have such a large fraction of people interested in astronomy through public education, but it should be taught through formal education in schools, where nearly all pupils study.

Frequently I have heard good amateur astronomers and schoolteachers interested in astronomy say that one should watch stars because of their beauty. It may be true for them and some pupils, but there are people, on one hand, who love stars, as well as people, on the other hand, who love the beauty of flowers. What we should understand is that it is a matter of

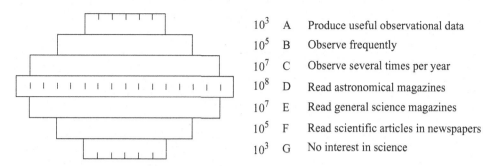

10^3	A	Produce useful observational data
10^5	B	Observe frequently
10^7	C	Observe several times per year
10^8	D	Read astronomical magazines
10^7	E	Read general science magazines
10^5	F	Read scientific articles in newspapers
10^3	G	No interest in science

Fig. 21.4. Same as for Fig. 21.1, but with a different structure. This structure is the ideal one.

preference for individual people to say, "I love and/or enjoy watching stars or flowers," but it is not real understanding of sciences such as astronomy or botany. If one can be trained to understand sciences – especially physical sciences – one often has an ability to understand and evaluate environmental issues.

21.3   In what way should astronomy be taught?

Countries fall into several types, as far as the teaching of astronomy is concerned. In some countries, astronomy is a compulsory topic in the school curriculum. In other countries, pupils can choose from physics, chemistry, biology, and earth science. In the latter case, only a small fraction of pupils (less than 10 per cent) choose earth science, which includes astronomy. A third way is to teach integrated sciences, by considering that daily phenomena cannot be explained without combining different kinds of sciences.

In order to go the third way, I proposed to teach "stories" connecting all the related sciences, depending on the pupils' grade and ability (Isobe, 2000). A set of examples is shown in Table 21.4. To proceed in this way successfully, it is important to include a number of well-prepared exercises. Otherwise pupils just listen to the teacher's stories but can seldom catch the stories' scientific goals.

As examples, we developed two exercises; the first one is evaluation of light pollution and loss (Isobe and Hamamura, 2001) and the second one is asteroid detection software (Isobe *et al.*, 2002).

One other important issue to be carefully considered is that the development of the sciences, especially astronomy, is very rapid. Therefore, depending on new discoveries and also on their exposition in newspapers and so forth, pupils' interest is changeable and we have to develop new stories continuously. This is somewhat demanding work, but is inevitable work for future generations.

There are different national and international studies, by such organizations as the United Nations and Organization of Economic Cooperation and Development, to compare pupils' ability – especially in physics and other sciences – in different countries. However, nearly all the countries that participate use the output to improve their ranking, and try to introduce an educational system carried out in countries with high ranking, without deep consideration of differences of culture, history, and nationality.

Table 21.4. *Some examples of stories containing items of physics, chemistry, biology, and earth science*

Topics		Physics	Chemistry	Biology	Earth science
Quality of material and amount	Multi-phases of materials	Effect of T and P	Chemical reaction	Growth of animal	Phase change of stone
	Feature of Earth	Effect of gravity	Compositions of atmosphere	Necessary condition of life	Earth is one of the celestial bodies
Funda-mental	Circulation of atoms	Nuclear fusion	Chemical character of each atom	Circulation of atoms through living species	Formation of stars and sun
	Solar radiation	Light spectrum	Chemical reaction for materials	Light effect to life	Climate change
Nature	Water flow	Evaporation	Chemical inclusion	Water for life	Erosion
	Dinosaurs	Area and volume	Explosion energy	Evolution of Life	Asteroids

21.4   Conclusion

Education systems are different from country to country. Even if a good education system is introduced in a country, its real results will become apparent only after several decades. Therefore, at all times, we have to communicate with each other closely. I hope that this special session will be a "trigger" and that some participants will try to set up a new Program Group of Commission 46 in order to communicate productively with other educational organizations.

References

Isobe, S. 1991, "Proposed structure of education in astronomy," *Publications of the Astronomical Society of Australia*, **9**, 72–5.

Isobe, S. 2000, "Report on International Conference on Primary School Science and Mathematics Education," held in Beijing, November 1–4, 2000, and organized by the International Council of Scientific Unions (ICSU).

Isobe, S. and Hamamura, S. 2001, "Educating the public about light pollution," in R. J. Cohen and W. T. Sullivan III, eds., *Preserving the Astronomical Sky*, IAU Symposium No. 196, International Astronomical Union, Dordrecht: Kluwer, 363–8.

Isobe S., Asami, A., Asher, D., Fuse, T., Hashimoto, N., Nakano, S., Nishiyama, K., Oshima, Y., Takahashi, N., Terazono, J., Umehara, H., Urata, T., and Yoshikawa, M., 2002, "Educational program of Japan Spaceguard Association using asteroid search – Spaceguard Detective Agency," in B. Warmbein, ed., *Proceedings of Asteroids, Comets, Meteors*, ACM 2002, held in Technical University Berlin, Berlin, Germany (ESA-SP-500), 817–19.

Part **X**

Conclusions

Closing discussion

Harry Shipman: Virtually everything that has happened in the conference is relevant to one or more of the recommendations of the resolution (see pp. 2–3). There should be a session like this at the next IAU General Assembly meeting and, preferably, it should not be on the last day.

Jay Pasachoff: I agree. Perhaps we are scheduled at the end of the meeting because we requested to be adjacent to the Teachers' Day. Next time we can ask to be earlier, and we can also ask for more than one session. Within the structure of Commission 46 on Astronomy Education and Development, we will certainly ask all our national liaisons to circulate the resolution within their respective countries. Further, we can ask the IAU national representatives to bring this resolution to a higher level: to the respective education ministries.

Case Rijsdijk: We should try to focus on bringing astronomy into the teacher-training institutions.

Syuzo Isobe: This resolution is a natural result of the present condition of astronomy education worldwide. However, although many IAU Commissions talk about the importance of education, the reality is that most of them only think about education at the research level, but not at the "mass" level. Therefore, our resolution should include a statement – backed up by IAU members, the IAU itself, and the national committees – to try to take school education seriously. This resolution could be presented to the UN, to the OECD, and to different organizations in a timely manner.

David McKinnon: Dictation to government, universities and teacher-training institutions (TTIs) is problematic. The government defines curriculum, to which TTIs and universities in part respond. Targeting universities and TTIs directly will not work. Also, we need to post the resolution on the IAU website. (*Editors' Note*: Done.)

Claus Madsen: On the question on what the IAU can do in support of the resolution, perhaps the IAU could develop a mechanism by which it could formally endorse projects and proposals at the national level. By imparting its scientific authority, it could provide valuable help to those who strive to improve science education in the countries.

Abdul Athem Al-Sabti: I think a letter should be sent from either the General Secretary or the president of the IAU to every minister of education. Other UN bodies should also endorse the letter. Although it requires effort to find out 150 names, it will produce results in the end, particularly in developing countries.

Ruth Ernest: It is unanimously agreed that education in astronomy is essential and generally accepted that educating teachers is an important part of this. I recommend that the IAU look at the possibility of setting up a liaison with classroom teacher representatives to work as a facilitator to aid in the achievement of the outcomes of the resolution.

Leonarda Fucili: I consider recommendation No. 3 very important in order to promote astronomy in national educational systems, but I'm asking myself how many national representatives are present in this symposium and are involved in education. I consider their involvement essential, and I propose an "educational commission" with one or two more active national members in any country.

Jay Pasachoff: We are honored to have the president of the American Astronomical Society in attendance – evidence of the increased interest in education in the society. I wonder if she would like to make any remarks.

Catherine Pilachowski: Our national astronomy organizations can play an important role to connect astronomers and teachers, and to convey the important goals of the IAU resolution. The American Astronomical Society has an active and ongoing education program, and I know this is the case in many other countries represented here. Our national societies already have the contacts and connections to be useful. For me, I see an important role for the commission to help us all become more informed about the initiatives and activities of other countries around the world.

Nahide Craig: Can the IAU include (through the US national liaison) in their list: the National Science Teachers Association (NSTA); American Association of Physics Teachers (AAPT); National Council of Teachers of Math (NCTM); and/or provide some funding for their science education leaders to participate? These associations lead the teaching of science and mathematics and make policies on the science book selections. Including them in these meetings or at least informing them about them will be useful for the IAU.

John Percy: This is a good idea, although there are not really any funds available. We will look into providing formal invitations to any relevant symposium at the 2006 IAU General Assembly or elsewhere; we will also notify these organizations about the availabilities of these and other conference proceedings.

Harry Shipman: As I think of the discussion of the past few minutes, I wonder how best to send a letter to the science education person in each place. I have known our director of science education, and her predecessors, for years. If she were to get a letter from John Percy, or Jay Pasachoff, or Ron Ekers, people whom she's never heard of, completely out of the blue, it would not be very effective. But if someone in Delaware – which presumably would be me – were to prepare the ground first, it could be much more effective.

Case Rijsdijk: I agree. I will approach our minister (or department) of education by working through high-profile people such as the head of the National Research Foundation. Astronomy has a high profile at present (including the 10-m South African Large Telescope, being completed in 2004 [SALT], and the bid for the Square Kilometer Array [SKA], a worldwide consortium to build a huge array of radio telescopes with that total collecting area), and the timing is right.

Moedji Raharto: It is necessary to send a letter from the president of this commission to the minister of education. I think also it is necessary to send a letter to the chairperson of the Indonesian Astronomical Association. It will take time to implement the resolution in the curriculum, and some conditioning is necessary to formulate what the problem is.

Anonymous: The national astronomical associations can run parallel sessions of education during their annual meeting. This will provide a very good opportunity for teachers connecting to professional astronomers. In Taiwan, we have the advantage that astronomy will be included in the test for all government employees.

Markiyan S. Chubey: Our suggestions should be seen as reasonable to introduce, as all our efforts in teaching astronomy and science in general are directed to saving and exploring the life of the beautiful Earth. It will aid the goals of ecological education, including the noble aims of science in general.

Conclusion

What is the future of astronomy education worldwide? How can astronomers and astronomy educators ensure that the future will be a positive one? A few years ago, one of us was invited to write an article on the topic of the first question (Percy, 1996). The article also addressed the second question, but it ended as follows:

There are alternate futures, of course. Young people may reject astronomy and other sciences as uninteresting and irrelevant, and turn to pseudoscience and fantasy instead. Teachers, journalists, and other opinion-formers may be ignorant and afraid of science. College enrollments in astronomy may shrink as students flock to easy courses, or to courses that promise more than the tired lecture approach. Planetariums may subsist on laser shows; science centers may evolve into elaborate video games. Astronomy may become an esoteric career or hobby, practiced by an eccentric but privileged few.

Fortunately, many of the characteristics of this "alternate future" do not appear to be developing – yet. But there are negative developments. Belief in pseudosciences, such as astrology, creationism, and space aliens, is widespread, and is not decreasing, as both Jayant Narlikar, and John Percy and Jay Pasachoff, have explained. Astronomical misconceptions, of many kinds, continue to abound. Astronomy is poorly taught, if it is taught at all, in many schools. The scientific enterprise appears to be maintained by a small fraction of the population. (It is therefore very important to nurture those "talented and gifted" students who are interested in astronomy; otherwise they may opt for a more "traditional" career such as medicine or business, or a field such as biotechnology that is perceived as being lucrative as well as exciting. We should heed the advice in Rob Hollow's chapter earlier in this book [p. 27].)

Another negative development is the relentless growth of light pollution. Light pollution is the unnecessary illumination of the night sky by human activity. It is a waste of energy. It is a symptom of inefficient lighting. It robs us and our students of the beauty of the night sky. IAU Commission 50 (Protection of Existing and Potential Observatory Sites) has an active education program; see: http://www.ctio.noao.edu/light_pollution/iau50/ and click on "education." Ironically, the study of light pollution is an excellent school topic that combines science, technology, society, and environment. There is an excellent teacher resource at: http://www.astrosociety.org/education/publications/tnl/44/lightpoll.html.

On the other hand, there are many reasons to be positive about astronomy education. Astronomy is advancing on all frontiers. There are exciting new results, with fundamental implications for the origin of the universe and its constituents, including life itself.

What can astronomers do to engage students, and the general public, in both the fundamentals and the frontiers of astronomy? By its nature, the IAU and its Commission 46

(Astronomy Education and Development) may not be the best organizations to lead this effort, since their focus is mainly on professional and university-level astronomy. The IAU is the only truly international astronomy organization, but it must work actively with other national and multinational organizations, such as the Astronomical Society of the Pacific and the European Association for Astronomy Education. Our European colleagues have been especially successful in organizing multinational partnership programs, through the European Union, as described by Claus Madsen in Chapter 19.

In principle, Commission 46's system of national liaisons could be of help. In practice, it may be necessary to augment that system by identifying, in each country, a handful of individuals who are *passionate* about astronomy education, and motivated to make use of international support, and knowledgeable about how to achieve results in their own country.

We must identify practical, relevant knowledge about *effective* teaching and learning of astronomy, and make that knowledge widely available in usable form. There is a great deal of useful information, but it is contained in a variety of sources. Bibliographies tend to be incomplete; it is interesting that the bibliographies in the chapters by John Broadfoot and Ian Ginns, and by Janelle Bailey and Timothy Slater, have little overlap. That's good in the sense that those of us from the northern hemisphere learned something new! A truly comprehensive, searchable bibliography is essential. Perhaps this would be a project for the *Astronomy Education Review*, described in Chapter 7 by Sidney Wolff and Andrew Fraknoi, or for the Astrophysics Data Service (despite its name).

A comprehensive bibliography, in itself, would not be of use to the average classroom teacher, or even to the average teacher educator. There is an additional need for a 20–30 page distillation of this information – a review paper on astronomy education, longer than this Conclusion, but shorter than this book. Some years ago, a group of North American astronomy educators noted the need for an authoritative review of astronomy education. They approached two respected astronomical publications which are known for the research review articles they publish. Both approaches were unsuccessful. Perhaps it is time for Commission 46 to develop such a review, to circulate it widely for comment, to publish it electronically, and to convince our colleagues to translate it if necessary and to disseminate it to the astronomy education communities in their countries.

Such a review might well have subsections. One could be on the ways in which astronomy has been *effectively* and *successfully* incorporated into the various parts of the curriculum (especially in science and, as eloquently described by Rosa M. Ros, in mathematics). Textbooks play an important role in defining what students are taught. This book contains chapters by Jay Pasachoff, a prolific writer of authoritative English-language textbooks at the school and university level, and by Jay White who, along with Don Wentzel, has assisted the Vietnamese astronomers in producing a single textbook in their own language. In many countries, there is a single textbook, often written as a "labor of love" by a local astronomer; Muhamed Muminovic describes one such book in his contributed abstract. Perhaps there is some way in which Commission 46 can support these efforts. Governments may be more willing to include astronomy in the curriculum, and to support astronomy education in general, if there is some reference to local culture and local astronomical achievements. The benefits of "local culture" are certainly apparent in the chapters by Julieta Fierro and Leonarda Fucili.

Education research would be another subsection. It is important to identify what areas of astronomy education research have been well covered, and what areas not. As noted by Bailey and Slater, and, Broadfoot and Ginns, solar-system topics have been well covered. But what about other topics? And there are issues other than content: what are the best approaches for developing students' skills and attitudes, and their understanding of the applications of astronomy? To answer these questions, of course, requires appropriate methods of *assessment* – both of students' learning, and of the teaching methods and materials. Remember that assessment includes *both* evaluation *and* feedback. And the feedback should lead to revision and improvement of the teaching methods and materials.

Then there is a need to disseminate information about effective, inexpensive astronomy activities and demonstrations. As Case Rijsdijk has pointed out, "inexpensive" has a different meaning for developing countries than for developed countries. In the former, it basically means "using available materials." At the other end of the spectrum: there was extensive discussion, in this book, of the Internet, remote/robotic telescopes, and related technologies. These are indeed powerful tools. We are glad that David McKinnon, in Chapter 9, discussed the assessment of these tools. There are obviously enormous benefits to be gained when these projects are led or advised by individuals, such as McKinnon, with expertise in astronomy, education, and education research. And let us not forget the advantages of "daytime astronomy" such as solar observing, radio astronomy, and data "mining" and analysis.

Another issue is "best practices" in teacher education, both pre-service and in-service, issues that were addressed by Michèle Gerbaldi and Mary Kay Hemenway. The solution would seem to be some basic training on good science pedagogy in general, plus training in the *basic* concepts of astronomy and astronomy teaching.

Finally, we must convince our colleagues in every country to consider investing at least one per cent of their time in astronomy education and outreach (the "one-per-cent solution," to use terminology related to Arthur Conan Doyle's Sherlock Holmes stories), and to convince their governments to provide corresponding funding to support their work. We must provide advice on how to achieve the highest impact with these limited resources (though low impact is acceptable if it has the beneficial effect of showing our colleagues that outreach is interesting and satisfying).

One solution is to make use of the skill and enthusiasm of amateur astronomers. In most countries, the number of serious or "master" amateur astronomers is comparable to the number of professional astronomers. In addition, there are ten times as many "journeyman" amateurs who would be willing and able to contribute to outreach. In the USA, the Astronomical Society of the Pacific and the Astronomical League (the national organization of astronomy clubs in the USA) have recently joined forces in a new outreach initiative. The ASP provides much of the basic educational material; the amateur astronomers present it in schools and to the public. Unfortunately, amateur–professional relations are quite cool in many countries, especially the less developed ones.

Partnerships are essential. In addition to professional–amateur partnerships, there need to be partnerships between professionals and the planetariums, and among astronomy educators, education researchers, and teacher educators. Commission 46 should have Program Groups (see Syuzo Isobe's chapter) in each of these three areas. It is especially important to link with the community of planetariums, science centers, and public observatories. Millions of people (including students) visit such facilities each year. Nahide Craig and Isabel Hawkins, as well as Nicholas Lomb, in their chapters in this book, describe the variety of important, high-impact programs that these facilities can provide.

How can we find volunteer members for new Program Groups? The IAU has recently reorganized itself into divisions, and Commission 46 is now part of Division 12 – Union-wide

Activities. One of the purposes of the new structure is to involve IAU members more directly in union activities. There are hundreds of IAU members with more than a casual interest in education and outreach. We must find ways of involving them in Commission 46 activities.

It is especially important to involve as many countries as possible in the work of the commission. At the conference on which this book is based, countries such as China and India were under-represented, as were many other countries in Asia, Africa, and South America. China and India have a long tradition of astronomical observation and research, and both countries are actively developing their space programs. There is great potential to involve their students and teachers in astronomy. The IAU has been active in astronomical development in many countries. Sad to say, astronomy is still struggling in most of them: Central and South American countries such as Paraguay and Peru, for instance. Countries such as these need follow-up support. This need not be financial; the IAU budget is limited. There are other countries in which there are few, if any, professional astronomers, but there is interest in astronomy in the schools, or among the public. The situation in the Caribbean is described in one of the contributed papers in this book.

There is also a need to continue the series of international conferences on astronomy education. The IAU has sponsored over 400 formal conferences, symposia or colloquia. Only two have dealt with astronomy education; the conference on which this book is based was a "special session" within a General Assembly. The IAU Executive Committee has been generally supportive of astronomy education, but there are special problems in organizing an education conference: for a science research conference, many participants – especially from the developed countries – can support their attendance through research grants; for an education conference, very few participants can. Extra funding is needed.

Funding is sometimes available through organizations such as ICSU and UNESCO. The problem is that the availability of this funding is not always known, or is not known about until the last minute. Multinational corporations and foundations might be another source of funding. Furthermore, it is important for such a conference to be effective. Certainly conferences are good professional development for those who attend. But there should be concrete outcomes and deliverables, such as the bibliographies, reviews, and resource lists described above.

What should be our goal for this decade? It would be unreasonable to expect to solve every problem of astronomy education, and to make every teacher and student an expert in astronomy. Syuzo Isobe, in Chapter 21, defined seven categories of interest in astronomy in Japan, ranging from a hundred people who produce useful astronomical data, to a hundred *million* people who have no interest in science at all. Our goal should be to move everyone upward by one category in Isobe's scale!

Reference
Percy, J. R. 1996, "The future of astronomy education," *Mercury* **25**(1), 34–6.

Author index

Subject index